KEY THINKERS IN RELIGION AND ENVIRONMENT

Key Thinkers in Religion and Environment provides a theoretical foundation for scholarship related to the intersection of religions, nature, and cultures across disciplines.

The text introduces students to the major names, theoretical issues, and methodological orientations of the field while giving professors maximum freedom to insert case studies and examples as they wish. Students will come away with an understanding of the most important scholars, their theoretical contributions, and the scholarly conundrums with which they wrestled. The book includes figures who are foundational to the field of religious studies more broadly, foregrounding key themes in their works which highlight the "nature" in/of their argumentation, whilst also highlighting the voices of women and people of color. The thinkers come from a range of fields, including religious studies, philosophy, anthropology, sociology, American Indian studies, ethology, agroecology, theology, and environmental history, demonstrating the importance and impact of interdisciplinary research. The book also offers a theoretical orientation which illuminates methodological and theoretical deficits in religious studies more generally, whilst opening new avenues for thinking about environmental ethics.

It is a must-read for all students and researchers of religion and the environment.

Lucas F. Johnston is Associate Professor of Religion and Environment at Wake Forest University, USA.

Interdisciplinary Studies in Religion

Key Thinkers in Religion and Environment
Theoretical Foundations
Lucas F. Johnston

KEY THINKERS IN RELIGION AND ENVIRONMENT

Theoretical Foundations

Lucas F. Johnston

Routledge
Taylor & Francis Group
LONDON AND NEW YORK

Designed cover image: Getty Images

First published 2025
by Routledge
4 Park Square, Milton Park, Abingdon, Oxon OX14 4RN

and by Routledge
605 Third Avenue, New York, NY 10158

Routledge is an imprint of the Taylor & Francis Group, an informa business

British Library Cataloguing-in-Publication Data
A catalogue record for this book is available from the British Library

Library of Congress Cataloging-in-Publication Data
Names: Johnston, Lucas F., author.
Title: Key thinkers in religion and environment : theoretical foundations / Lucas F. Johnston.
Description: 1. | Abingdon, Oxon : Routledge, 2024. | Includes bibliographical references.
Identifiers: LCCN 2024001627 | ISBN 9781032428970 (hardback) | ISBN 9781032428987 (paperback) | ISBN 9781003364801 (ebook)
Subjects: LCSH: Nature--Religious aspects. | Human ecology--Religious aspects. | Human ecology--Moral and ethical aspects. | Environmental ethics.
Classification: LCC BL65.N35 J64 2024 | DDC 201/.77--dc23/eng/20240315
LC record available at https://lccn.loc.gov/2024001627

ISBN: 978-1-032-42897-0 (hbk)
ISBN: 978-1-032-42898-7 (pbk)
ISBN: 978-1-003-36480-1 (ebk)

DOI: 10.4324/9781003364801

Typeset in Sabon
by KnowledgeWorks Global Ltd.

CONTENTS

INTRODUCTION

Staking Out the Terrain

I was at a life crossroads, and I had gone into the woods to meditate on the possibilities that lay before me. I hiked a familiar trail, one with little traffic, stopping at the point furthest from the trailhead. I sat quietly, weighing my options, and enjoying the calls of the birds. I did not see it materialize, but when I turned my head slightly there was a bobcat uncomfortably close to me. It did not appear to be on alert, rather it simply sat, like a lazy hound on a summer day, looking directly at me. I was alarmed at first, for I did not see it approach. I gazed back at the creature curiously, knowing that they are typically wary of and avoid humans. I wondered, half whimsically, whether the bobcat was an emissary of some kind, acting as a sort of guide to my decision-making. And then, as quickly as it had appeared, it was gone. I was not even conscious of shifting my gaze from it, but I must have, I reasoned, for I did not see it go any more than I had seen it come. Under the circumstances, the unusual behavior of the bobcat during a weighty decision-making process occupied my thoughts, and I realized that I really *wanted* that bobcat to be some kind of mysterious messenger from the spirit world. Whether it was, in fact, was immaterial. The cat, unfortunately, did not divulge any secrets for steering my life choices. But the episode had motivated me to consider more carefully my relationships with and awareness of animals.

I had relationships with domestic non-human animals for all my life, and it was easy for me to imagine them as agents in the world—aware of their own needs and acting purposively to fulfill them. It was more difficult perhaps to get into the mind of a wild animal, I thought. But over time I developed a perception that things like bobcats, bears, bats, and bonobos (and more) might

DOI: 10.4324/9781003364801-1

have their own "good," and that they might actively work toward the fulfillment of that good. Perhaps like my domesticated companions, these "others" were worthy of respect, and should be approached with a certain etiquette. I did not then have a name for those sentiments. But such a sensibility was one elaborated by many of the scholars who feature in this book. Some early scholars who attempted to study religions from an empirical perspective recognized that animals, features of the landscape, and natural phenomena were intimately tied up with the religious lives of some cultural groups. For example, E.B. Tylor, Emile Durkheim, and Mircea Eliade looked for commonalities in religious experiences and social formations where expanded notions of personhood were imagined as *primitive* thinking (see Chapters 1, 2, and 3). Jane Goodall and others featured herein who advocated for the personhood of other-than-human animals posited that commitments to others (whether ethnic, cultural, ethical, or intraspecies others) are useful for navigating the worlds that humans find themselves thrown into. Many of the thinkers who star in this book were not themselves adherents of institutionalized religious organizations. But they all had something important to say about the intersection of religions, nature, and social formations. My bobcat experience, for just one example, would not "count" as religious or spiritual for some who view practices, experiences, and beliefs as confined within institutional settings. But for me, it was formative, a flashbulb memory accentuated with feelings of curiosity, fleeting fear, and a growing comfort and belongingness that ran in and through my body following the experience. Imagining that sort of personal, embodied experience is at the core of religion, however, is to miss the social categories that are always in play with socio-political deployments of the term religion or spirituality. *Nothing is religious that is not already, always political.* Each use of the term religion opens the door a bit to give us a peek at what *others* mean when *they* use such terms.

In the twenty-first century, the number of people who identify as "none of the above" religions when participating in interviews and surveys became the largest single constituency in the United States, overtaking evangelical Christianity, which held the title for decades. The category of "nones," as they were often called, was a basket for catching lots of religion-resembling cultural production. But for some time, there was little work disaggregating the nones category—scholars delayed in making finer-grained distinctions between those who occupied that relatively vague category. Importantly, many of those who defected from religions did not wish to completely jettison any sense that there was a higher power of some kind. They did not wish to identify as agnostics or atheists. But they also did not wish to be identified as religious. In the late twentieth century and into the twenty-first century, many people began to use the term *spiritual* specifically to set themselves apart from those who identified as religious. The content of the term spiritual, because it became widely used, was variable. In general, however, it can be said that spirituality was often imagined as a less formal, more personal sensibility than religion (Roof 1993: 76–77; Taylor 2005a). Spirituality was related to unique life experiences, moral values, and sensibilities; religion was organized and required collective adherence. As the scholar of religions Bron Taylor put it, "Nowhere is the preference for the term 'spirituality' over 'religion' more prevalent than among those engaged in nature-based or nature focused religion" (Taylor 2005a: viii).

For analytical purposes here, religion is imagined as a broader umbrella category, including institutionalized forms, as well as more diffuse and personal senses of spiritual seeking. Importantly, there is no essential difference between religion and spirituality. However, the individuals and groups who deploy the term in the public sphere mean different things when *they* use the terms. Religion is a notoriously slippery term, and what religion *is* depends upon the person using the category and what they gain by using the term. For example, some environmentalists have described environmental movements in religious terms, hoping that the emotive content of religions would complement scientific descriptions of the environmental crisis and motivate beneficial action. Others have referred to environmental movements as religious to discredit them. They cast environmentalism as a dangerous, misanthropic (anti-human) pagan religion which hurts poor people. So, religion is not some*thing* that exists "out there" that the critical cultural detective can point to and say, "ah, there's religion!" Rather, as one scholar of religions lamented, he was able to find more than 50 different definitions of the term. Relating this story, the scholar of religions J.Z. Smith countered that this did not mean that it was impossible to define the term. It meant that the term could be productively defined in at least 50 different ways! In other words, each of those definitions were products of the scholars using them, and the questions they hoped to illuminate in their investigations. Religion was not the independent variable driving human behavior. Instead, it was a variable dependent upon the shape of societies and the ways in which power could be exercised within those societies. The category religion was a sharp tool which, when doing public work, cut out political, economic, and social differences.

If religion and spirituality have no essential core that differentiates them, if they are tools for social formation, then related terms such as *sacred*, and *holy* have similar characteristics. The sacred is not some quality abstracted from the people, places, or things it is applied to. Rather it is a way to publicly cordon off things constructed as sacred and separate them from what is supposedly profane. The use of such terms does not render objects *authentically* sacred, or holy, or what have you. Instead, such terms are useful to specific individuals or social groups expressing their own perceptions of the way things ought to be. To utilize an example from the anthropologist Mary Douglas (1966), which has been employed in introductory religious studies texts (McCutcheon 2019), the difference between soil and dirt is a product of social classifications. Likewise, distinctions such as "sacred" and "profane" are not inherently obvious or meaningful. They become so because they are authorized by context-dependent webs of meaning and specific power arrangements. Religion and the sacred (and associated terms) should not be imagined in an essentialist way, as fixed things, or phenomena, but as a function of the imaginative acts, interests, and values of the individual or group using the term. The task of the cultural critic is to approach all uses of such terms with suspicion, focusing on their observable political, social, and economic impacts. In other words, their social function. They must analyze *which* terms are being exercised, by *whom*, and with what effect (Martin 2023; McCutcheon 2019).

When thinking about religions and environments,[1] there are several ways in which those two categories might be related. To attend to the contributions of these key thinkers, I refer throughout to *environmental religions*. The content of this term is unpacked more below, but it can refer to:

- Cultures which associated natural elements with deities or perceived them to be inspirited in some way.
- Existing religions which were re-interpreted through the lens of current ecological crises.
- Religions which included cosmological narratives with ethical implications, limiting human behaviors relative to the environment.
- Religions which supported social justice causes with distinctive environmental components.
- Groups that were not themselves religious, but which deployed religious or spiritual language, metaphors, or symbols drawn from nature.

Social justice, noted above, is the notion that society is both unequal and inequitable and that activism may create more just social arrangements. Equality is the idea that all individuals ought to have the same resources and opportunity structures. Equity refers to fairness. Providing an individual with meager means, and an upper-class individual with the same set of resources, is an exercise in equality. Acknowledging that their social contexts facilitate different levels of access to the profitable use of those resources is a recognition of inequity. Much social justice advocacy involves instances of environmental racism. Environmental racism refers to instances where the health impacts of ecological degradation or industrial pollution are endured by minoritized communities. In the United Sates, a dramatic majority of health-damaging pollution or waste storage occurs in communities of color. Social justice is an important feature of some environmental religions. To sum up, I think it is less important to provide a concrete definition of environmental religion than to offer a skeletal characterization of term and illustrative examples.

Many environmental religions emerged in response to the fraying of human relationships with the environment. And many attempted to restore humans to their rightful place as part of the ecological system upon which they depended. Environmental religions are not, then, just nature religions (those grounded in reverence for natural features and creatures), nor are they encompassed by efforts of global religions to "go green." There are also secular environmental movements with significant religious dimensions.

To explore environmental religions, it is not necessary to review the many perspectives on the historical relationships between religions and environmental consciousness in North America and Europe. That task was accomplished by a bevy of others upon whose work I draw (Berry 2015; Glacken 1967; Johnston & Taylor 2016; Kearns & Bauman 2024; Marsh 1864; Nash 1967; Stoll 2015, 2023; Taylor & Johnston 2016; Worster 1977). The emergence of ecological restoration and conservation biology in the 1980s were examples of action-oriented, applied sciences that viewed conservation and restoration as a sort of penance for ecological destruction (Elliot 1997; Jordan 2003, 2005; Jordan & Lubick 2011; Johnston 2013: 78–103; Katz 2003, 2012; Light & Higgs 1996;). Examples of environmental religions derived from the human or natural sciences were potent examples of the spiritual dimensions of scientific research. Environmental religion is meant to be a broad term referring to all religious cultural production in which nature, however conceived, plays a role. Environmental religions may be dark green, or light green, or may not even appear very green at all. Environmental religions, like any religions, have elements of cosmology and ethics, and include different understandings of reality, or different knowledge systems. The study of the question of what is *really* real, is ontology—there are different understandings of reality. The study of knowledge systems, how knowledge is gained from the world and how it is interpreted, is called epistemology. What counts as important data about the world? The answer varies in different cultures—they have different epistemologies. These terms (ontology and epistemology) will appear throughout the book and are important components of understanding environmental religions.

To dodge some of the troubles with defining the category religion, the scholar of religions David Chidester referred simply to things that do *religious work*. Something was doing religious work, he said, when it bound communities of people together, focused desire and moral vision, and channeled exchange relations. Religious work was observable in social action and its effects. A short list of some of the ways in which environmental religions are doing religious work may be helpful (although it is not intended to be exhaustive). There are:

1 Religions which, from their founding, articulated or endorsed the care for and protection of the environment. The Baha'i religion is one example,

where adherents focused on the interconnection of humans and their ecosystem homes.[2]

2 Secular environment- or sustainability-oriented social movements that drew on religious metaphors and tropes to advance their causes.
3 A growing number of animistic and paganistic spiritualties, whose sources of inspiration and practice were natural features of the environment, or inspirited places. Animism refers to the notion that specific features of the landscape, or non-human animals, possess a spirit which animates them (see Chapter 1). Paganism is a general term used to describe religions that pre-dated the emergence of Christianity, and also in some cases the contemporary resuscitation of ancient practices in Neo-paganism.
4 Western scientific narratives were often imagined as promoting a sense of awe or ultimate dependence on the energetic systems of which we are a part. These included taking scientific narratives themselves as sacred stories applicable to all people.
5 Empirical methods utilized by those outside of Western worldviews whose narratives perform the same functions that Western science plays in the global North—that is, they are exploratory, and explanatory narratives. Traditional ecological knowledge (TEK) is one example of this sort of environmental religion.
6 Some religious leaders and groups have endeavored to excavate their religious traditions and reinterpret them as having some environmental wisdom relevant in the present.
7 Religious individuals and groups that acknowledged that their texts were written in a time when environmental concern, as we know it today, was not present. But such groups might approach environmental issues through the lens of social justice.

All of these "count" as environmental religions. The ontologies and epistemologies of various environmental religions are data for analysis. The cultural critic does not need to inquire as to whether something was *really* sacred, or whether the image of a face on a piece of toast was *really* placed there by a divine being, as some individuals claimed in a popular and widespread preoccupation with seeing religious figures in food. The observable political, social, and economic impacts of claims about the sacred or holy are the object of study, for they are still true whether or not some nonempirical being is invoked.

Each of these above modes of environmental religion makes an appearance in this book. Some of the categories overlap, for instance, where some institutionalized religions engage in both mining their traditions for glimmers of green (number 6 above) and promoting environmentally sustainable practices (number 7). In other cases, there are contestations between them, for example when TEK (5) is deployed as a corrective to Western science (4).

To further situate this book, it is necessary to acknowledge that this is not *the* definitive list of scholars contributing to scholarship on religions and environment, but one aimed at Anglophone audiences and studies at the intersection of societies and ecosystems. People from other geographic locations (i.e., from East Asia, Africa, or other parts of the world) would certainly produce different lists of featured scholars. The ones in this book are meant to illustrate the breadth of disciplines that shaped research on environmental religions in the Anglophone world. Many of those included do *not* fall under the umbrella of institutionalized world religions. The classic scholars Edward B. Tylor, Emile Durkheim, and Mircea Eliade are included not because they were themselves religious (Tylor and Durkheim, at least, were atheists). Rather they studied people who, through ritualized performance, tried to remain in right relations with their habitats, characterizing them as backward and primitive (Malinowski 2013, 2020; Muller 2002; Tylor 1883; Frazer 2012 [1890]). There are other sources where readers can encounter important outlines of those scholars' research. But here we will read these classic scholars *through* an environmental lens, noting how their research programs portray the relationships between people and environments. The remaining chapters analyze 20th and 21st century figures and the modes of environmental religion they deployed.

This book is intended to be useful as a text for upper-level undergraduate courses on religion and ecology, nature religions, environmental ethics, and related areas of study, and for introductory graduate student courses in religion and environment. Through the cast of characters, readers will become at least somewhat familiar with many major avenues of research on religions and environments, will gain an interdisciplinary look at how different scholars historically understood and explained these complex social relations, and get some sense of ongoing research programs which are related to or derived from these foundational figures.

The disciplinary habitats of the featured scholars include anthropology, sociology, religious studies, environmental philosophy, theology, American Indian religions and law, environmental history, arboriculture, agroecology, and biology. All of them are important foundational thinkers for scholarship at the intersection of religions and environments. This book is a basic introduction to the study of environmental religions. As cultural critics, it might be useful to know the general contours of Durkheim's social theory, any signposts to look out for in the terrain of environmental philosophy, or the topographies of power, politics, and justice wrapped up in the preservation and/or sharing of TEK, for a few examples. For graduate students, these are the people one needs to understand to hold theoretical conversations with future colleagues. For undergraduates, these thinkers are figureheads who have made important shifts in how we think about, and relate to, nature.

Communities of Accountability and Environmental Social Movements

Whoever controls the access to and the distribution of resources (natural, social, and financial capital) rules. Both social and natural capital ultimately depend on the continuing health and viability of Earth, whose only real external input is sunlight. Otherwise, it is for all intents and purposes a closed system. That singular input of sunlight is modified and transmuted into the ability to do work through the food web, from producers, to consumers, to decomposers, all of them contained within the planet's finite system. This is our home. It is from the Greek term *oikos,* or home, that terms important to our study derived. *Oikos* is the root term for two important English words, *eco*logy and *eco*nomy. Economics is the study of the exchange relations in "the house." Ecology is the study of all relationships (not just exchange relations) that exist in the "home." When it comes to policymaking, however, most people imagine the economy as the most important metric. In real life, any economic activity depends upon the resources that sustain it: economics is a subsystem of the ecological system.

Economists sometimes assume that any limits in the system can be pushed back with technology. But even if limits can bend, they are still there. At this writing, the top 20 countries in the world with the highest fertility rate are in Africa, and at least half of them are already conflict zones (Thiede et al. 2020). Because these conflicts are at bottom about scarce resources, they are likely to worsen as higher populations compete for rapidly liquidated resources (Sciubba 2021; Smil 2008, 2010; Klare 2008, 2012). So, there are growing numbers of people born into war-torn geographies, which means more conflict and suffering.[3] These trends have moral as well as demographic implications.

The anthropologist Scott Atran investigated forms of *imagined kinship* that went beyond familial kin to identification with larger, amorphous, transnational groups of people characterized by their religious practice. So, for example, some members of a Muslim community in North Africa felt some sympathy for and even took action to support the struggles of other Muslims (i.e., Palestinians or Afghans) against Western powers. In such cases, religion was used to justify, or explain resistance to colonial rule, "development," and the usurpation of control over resources and land (Atran 2002; 2005; 2011; Atran & Medin 2008; cf. Taylor 1995). It was a sort of social glue for cross-cultural, transnational identities and ideologies of resistance. *Identity fusion* refers to the ways that perceived affinities and alliances emerge from material exchanges and shared desires (Atran et al. 2015). Religion is one of the most powerful ingredients of identity fusion. These two elements, imagined kinship and identity fusion, tied to sacred values, create a positive feedback loop, each strengthening the other and creating circumstances in which the choices that individual actors make are not reducible to "individualistic" rationality. I refer to those imagined kin, those with whom we identify, as communities of accountability.

All persons belong to lots of different communities of accountability. That is, everybody has multiple other individuals, groups, or relationships which, taken together, comprise their identity. That identity is expressed in performative acts of living, and which of those communities is most formative, or influential in our decision-making might change over time. We are more accountable to some of these communities or relationships than others. For example, one's family is a community. As is a soccer team, or a band, or a chess club—all communities. Other affinities are relevant also, for instance, Texans are a category, and citizens of nations are another category. One can even imagine being accountable to a global community.

Green and Brown Environmental Social Movements

In thinking specifically about religions and environments, it is important to imagine how they might manifest in social movements. The scholar of religions Bron Taylor characterized what he termed *dark green religions*, those that evidenced some sort of reverence for life ethic, grounded in either a bio-centric (life-centered) or eco-centric (ecosystem-centered) ethic (see Chapter 8), and a spiritual sensibility that accompanied perceptions of the sacredness of nature and/or its inhabitants.[4] In his analysis, Taylor offered a graphical representation ranging from naturalistic to supernaturalistic orientations (graphed on the y-axis), and from "Gaia" spirituality to animistic spirituality (on the x-axis). The Gaia hypothesis was pioneered by James Lovelock and Lynn Margulis. The term *Gaia* was an ancient term for the Earth goddess from Greek antiquity. Lovelock and Margulis hypothesized that the Earth itself was a self-sustaining "organism," that its systems fostered a sort of planetary homeostasis. If the Earth itself was a self-regulating system and humans defrayed ecological and social systems, then Gaia herself would eventually restore equilibrium, whether it included humans or not. On the x-axis, then, there were animistic perceptions that the natural world or elements of it were inspirited (animistic), and perceptions that there was something more, some higher power (i.e., the Gaia hypothesis). On the y-axis are people who have a deep reverence for scientific descriptions of nature without the need for spiritualized language and metaphor (naturalistic). On the other hand, some dark green religious practitioners also perceived some sort of preternatural or supernatural element that animated their dark green perception. The supernaturalism was rendered in different ways, from monotheistic positions, to the perception that forces or spirits actively impacted human affairs.

The environmentalist milieu does contain distinctly *dark green* elements. But in some cases, environmental religions do not appear to be very green after all (Johnston 2013). For example, many environmental advocates are not themselves outdoor enthusiasts and would not immediately or easily connect their understanding of nature to their religious perceptions. Yet some of them would unhesitatingly use highly emotional, ethically charged, or religious language to advocate nature preservation if it promoted their vision of environmental sustainability. Imagine the different constituencies that fall under the umbrella of environmental social movements as a series of concentric circles, moving from a *dark green* center, to *light green*, to *light brown*, to *dark brown* in the outside circle. The central dark green circle represents those who may resonate with some form of nature-as-sacred religion, or ideas about the intrinsic value of nature.

Light green environmental religions recognize some correlation between social justice and ecological degradation, but many individuals may not agree with the notion that nature itself is sacred or has intrinsic value. This light green group could be characterized as being anthropocentric—that is, human-centered. This includes activists who are interested in or advocates of environmental care who came to the environmental cause through other avenues. For instance, many environmental justice advocates discerned that one clear manifestation of injustice was the disproportionate health impacts of environmental degradation on minoritized or marginalized communities.

There are also *light brown environmental religions*, where environmental issues were a subset of concerns related to racial, ethnic, ethical others who were the targets of all manner of injustices. Environmental health and wellness are important to the extent that they provide for the flourishing of marginalized populations. The presence of religious ideas, metaphors, tropes, and argumentation with environmental elements have been prominent in communities of color since the 1970s. In the light brown constituency are individuals or groups that do not focus on environmental issues at all, but still work toward social equity, justice, a modified and sustainable mode of capitalism, or perhaps the overturn of capitalism in general. They critiqued environmentalism as a Western concern and noted that its politics required the dispossession of land for the creation of "wild" spaces. Imagine that these constituencies represented a *light brown* sentiment, one that focused primarily on social engineering for the greater good, with environmental protection and care a desirable by-product. We might also include, for example, critiques from environmentally-minded scholars in the global South who decried the ethnocentric suppositions at the root of conservation and preservation movements (Denevan 1992, 2011; Glass 2005; Guha 1985, 2000; Spence 1999).

Those outside the environmental milieu dwell in the *dark brown* circle, which represents those who typically do not support the central aims of environmental social movements or have affinities for those who reside within the other portions of the diagram (Johnston 2013: 32). They may perceive that

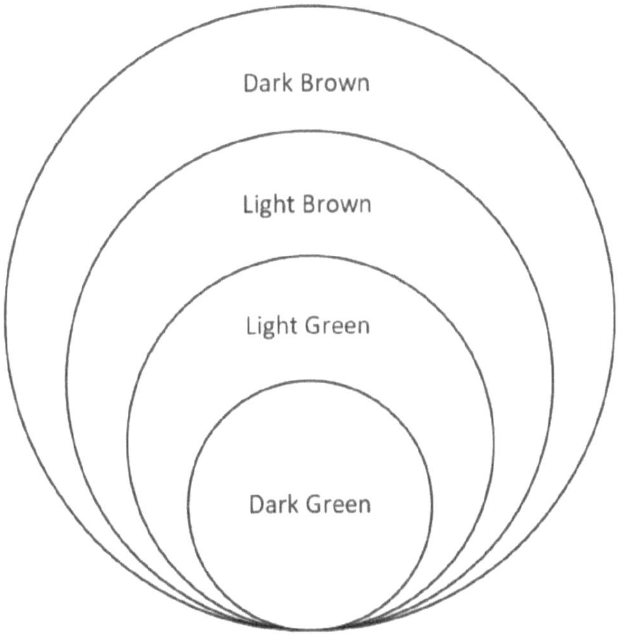

FIGURE 0.1 Types of environmental concern.

environmental causes are inherently misanthropic and that politics guided by such causes are dangerous. Much like Taylor's typology, each of these concentric circles may contain supernatural, preternatural, or naturalistic modes of explanation. Those communities of accountability all have distinct ways in which they prioritize the elements of their worldviews (Figure 0.1).

Environmental Religions and Global Politics

One of the ways that environmental religions have become especially visible is within international policy documents, programs, and publications. International attention to environmental concern began in the 1960s in North America and Europe (Johnston 2013), and by the 1970s international non-governmental organizations were affirming the obligation for Earth care or stewardship. In many cases, those statements were human-focused, with ecological factors included to the extent that they promoted human flourishing. They tacitly accepted human exceptionalism (humans were qualitatively different from the rest of the animal world) and advanced an anthropocentric, or human-centered, ethical stance. Humans might be drawn to care for the non-human environment because of the threat of their own demise (Schaefer 1970), or because there were in the scriptures of various traditions narratives that might promote care for ecological systems. Discourses related to *development* always, even if sometimes implicitly, included a human right to a healthy environment. Some examples will be illustrative.

Both the National Council of Churches (NCC) and the World Council of Churches (WCC) activated campaigns related the human and nature nexus in the 1960s. Importantly, in 1968 the United Nations Economic and Social Council (ECOSOC) made a resolution to tackle the "crisis of worldwide proportions...the crisis of the human environment" (Jolly et al. 2004: 125; Johnston 2013: 55). The momentum of these trends resulted in the United Nations Conference on the Human Environment in Stockholm, Sweden (1972). Poverty was imagined as the driving force behind human vulnerability to environmental fluctuations and over-exploitation of local resource bases. One particularly important outcome of this conference was the emergence of the phrase *sustainable development*, which included ecological relations as a concern, accounted for by the prevailing capitalist mode of production.

As the development expert Gilbert Rist argued, the notion of development was a key to the West's economic, political, and ideological conquests, for "by defining itself as the precursor of a history common to all, the West could treat colonization as a generous undertaking to 'help' more or less 'backward' societies along the road to civilization" (Rist 1997: 43). What Rist called the "global faith" of development and later sustainable development was increasingly touted by governments and multinational groups such as the United Nations (UN) as the means to alleviate poverty and increase security (Rist 1997).

Another important landmark was a series of publications predicting social collapse due to unpleasant and deadly encounters with ecological limits. The foundational book *Limits to Growth*, the subtitle of which referred to the "predicament of [hu]mankind," argued that if societies went on with business as usual, ecological limits would be eclipsed within 100 years (Meadows 1972). Paul Ehrlich's *The Population Bomb* (1968) associated those ecological limits with the population—the carrying capacity—of the planet. It was at that time particularly rare to invoke population as a cause of environmental degradation, so that book garnered varied responses both positive and negative. Economists decried these publications as alarmist and insisted that ecological limits could bend without breaking, and that the hand of the market could supervise the pressures on ecological limits. Ecologists, on the other hand, welcomed the conversation—finally (they believed) a work that implicated population as one of the primary drivers of ecological decline. Unsurprisingly, religious organizations were some of the groups that vociferously opposed any conversation regarding the limitation or management of human population. Particularly religions that traced their foundation back to the biblical figure Abraham were encouraged to multiply and control nature and were complicit in obscuring the role of population growth in environmental decline.

Garrett Hardin was one ecologist who put population at the forefront of his famous arguments in "The Tragedy of the Commons" (1968), and "Lifeboat Ethics" (1974). Hardin suggested that the influential notion of a "spaceship Earth," popularized by the economist William Boulding (1966), was useless because it posited a spaceship governed through cooperative goodwill and common goals. The lifeboat, said Hardin, was a more realistic metaphor. The spaceship would eventually list and crash because there was no captain. A ship without a captain does not get very far. What was necessary if humans were to emerge from the Anthropocene bottleneck alive, Hardin claimed, was "mutual coercion mutually agreed upon" presided over by an authoritarian captain.[5] The political scientist William Ophuls was even more explicit, arguing that only a coercive politics along with a "green religion" could prevent ecological collapse (Ophuls 1977).[6]

Other responses during the emerging environmental age included the convention of the "Program to Combat Racism" by the WCC in 1971. At the time development experts referred to the "Fourth World," populations in underdeveloped nations that were further disenfranchised by their own governments. This was clear evidence of a light brown environmental sentiment, highlighting the plight of Indigenous and traditional peoples throughout the globe. The anthropologist Robin Wright described Fourth Word societies this way:

> indigenous peoples throughout the world are united by their common situation as disenfranchised people, whose existence depends on a moral claim but who challenge the First World to examine its institutions, structure, values, which have left indigenous peoples powerless and dependent.
> *(Wright 1988: 373)*

The knowledge of Indigenous peoples, he noted, was crucial for the survival of ecosystems and its disappearance often disastrous. The Brandt Commission published two treatises on strengthening the relationships between the global North and the Global South. The first was *North-South: A Program for Survival* (1980), and the second *Common Crisis: North-South: Cooperation for World Recovery* (1983). Both foregrounded the now common notion that the benefits of "development" accrued to societies in the global North, overlooking the global South. Both were light brown in orientation, but certainly contained language of values and ethics at a global scale.

The 1983 World Commission on Environment and Development (WCED) was an important milestone, explicitly building on the Brandt Commission and advancing the idea that environmental degradation required international cooperation to protect environmental resources and heritages. Larger scale cooperation for environmental conservation was advocated in their resulting publication *Our Common Future* (WCED 1987), a publication that is light green throughout. While the scale of international cooperation recommended there is reminiscent of Hardin's and Ophul's arguments for global governance, the WCED did not recommend the coercive, authoritarian politics of Hardin and Ophuls. For many, the *Our Common Future* appeared to be good faith concern about the environment, but ironically imagined it manifesting in continued economic growth. As the Secretary-General of the Brundtland Commission stated, "the maxim for sustainable development is not 'limits to growth'; it is 'the growth of limits...many present limits can be expanded through changes in modes of decision-making, through changes in some domestic and international policies, and through massive investments in human and resource capital" (quoted in Smith 2005: 79; Johnston 2013: 63). Indeed, this attitude was characterized by the historian Donald Worster, who noted that development (and later sustainable development) allowed "the capitalist and the socialist, the scientist and the economist, the impoverished masses and the urban elites... [to] happily march together on a straight and easy path, if they did not ask any potentially divisive questions about where they were going" (Worster 1993: 144). The Joint Appeal by Religion and Science for the Environment explicitly called for a green religion: "Insofar as our peril arises from a neglect of moral values, human pride, and arrogance, inattention, greed, improvidence, and a penchant for the short term over the long, religion has an essential role to play. In so far as our peril arises from our ignorance of the intricate interconnectedness of nature, science has an essential role to play" (Johnston 2013: 65).

One specific outcome of the WCED was the recommendation for a global conference on environment and development (WCED 1987: 343). The result was the 1992 United Nations Conference on Environment and Development (UNCED) in Rio de Janeiro. There were several outcomes of that conference, including the approval of Agenda 21, a document of more than 800 pages which outlined guidelines for sustainable development with special attention to Indigenous issues.[7] One important document offered up at the conference was the Earth Charter, a clear statement of a global ethic grounded in a perceived kinship with the rest of the natural world (Taylor 2004: 1003). In this, the scholar of religions Bron Taylor noted, lay the emergence of a sort of global scale Earth religion (Johnston & Taylor 2016; Taylor 2010; Taylor & Johnston 2016). Some religious groups recognized this call for an environmental religion and took some action toward addressing the United Nations Sustainable Development Goals (SDGs).[8] From 1992 an organization called the Forum on Religion and Ecology (FORE) tended to its legacy and sought to spread its vision (Kung 2008; Tucker 2006, 2008).[9] But all-encompassing narratives were and are not the norm in much (if not most) of the world (Atran & Medin 2008; Johnston 2013; Sosis & Alcorta 2003). People from different social circumstances perceive different realities.

Civil Society Responses to Environmental Religions

Some civil society movements echoed or affirmed the global scale ethic of the Earth Charter. Others attempted to differentiate their struggles for sovereignty from global statements. They feared that such documents might be easily adapted into business-as-usual economic solutions which left those in the global South (the Fourth World) out of policy deliberations. More localized efforts still required a certain amount of awareness and encounter with cultural, ethnic, and ethical "others" (Johnston 2013: 71–74; 2014). Empathetic engagement, a process of personal or group cooperation, as well as a sense of kinship with the other-than-human lifeforms with whom we share our planet, has been a hallmark of both dark green and light green constituencies for decades. Many environmentalists envision their primary task as opposing the sort of global development that the multilateral development agencies like the World Bank, and international political regimes such as the United Nations, advertise.

The turn to Native societies and to the wisdom of Eastern traditions as counterweights to institutionalized sources of wisdom and authority was well-documented (Chapple 2000, 2002; Fields 1992; Girardot et al. 2001; Johnston 2013, 2014, Tucker & Berthrong 1998; Taylor & Johnston 2016; Tucker & Williams 1997). Romanticized renditions of Native and Eastern religions portrayed them as inherently environmentally ethical. The reality was more complicated, and historically both Native populations and Eastern ones sometimes caused the collapse of resource bases and thus societies (Diamond 2001; Tuan 1974; Krech 1999). But such characterizations of beneficent savages, or wise seekers of enlightenment, were frequently invoked as a part of Western European and North American environmental religions (Taylor & Johnston 2016: 356). Indigenous populations were long victimized by that sort of cultural appropriation. However, Native and traditional communities also mounted resistance to colonialist powers, especially in contestations over sacred lands. In the United States, for example, hundreds of Native peoples occupied the island of Alcatraz, once a famous prison in the San Francisco Bay, for more than a year. Networked supporters ensured supplies of crucial needs for the occupiers. They called themselves the "Indians of All Tribes" (Deloria 1992). Wounded Knee was another place which gained notoriety in 1893 as the place where over 200 Lakota women, children, and some elders were murdered by federal soldiers for the performance of Ghost Dance rituals, part of a millenarian religious movement (see Chapter 5 for more details). In 1973, the American Indian Movement (AIM) occupied that hallowed ground to call attention to Native land sovereignty. Violence erupted between Native activists and federal agents, and many were arrested.

Mount Graham in Arizona was a sacred landmark for nearby tribes, but also home to an endangered species, the Mt. Graham red squirrel (Taylor 1997). There was an opportunity for both conflict and cooperation as environmentalists and Native peoples worked to halt the construction of a large array of telescopes commissioned by the Vatican Observatory. Chi'chil Bildagoteel, Oak Flat, is a place sacred to the Apache in Arizona. Wendsler Nosie, Sr. was the most outspoken advocate of this Apache site, and he traveled domestically and internationally to draft more sympathizers to his cause. Another high-profile case involved the Dakota Access Pipeline (DAPL) which ran less than a half mile from the Standing Rock Sioux Reservation on land that was taken from them by the government in 1958. In 2020 and 2021, there was an uprising of "Waterkeepers" representing many tribes and other concerned peoples from across the world, to resist the development of those lands by fossil fuel companies (Johnson & Kraft 2017). These are but a few of the high-profile cases in which Native American environmental religion was exercised against governmental and corporate interests. All of those cases depended on the argument that the land itself was sacred, and that it must be ministered to by traditional peoples (Baker et al. 2017; Johnson & Kraft 2017; Peterson & LeVasseur 2017; Schneider-Mayerson 2015; Weatherdon 2017; Witt & Taylor 2017).

Among the most influential concepts and political philosophies that impacted countercultural milieus were Deep Ecology (Naess 1973), the Gaia hypothesis (Lovelock 1979, 1988, 2005), and bioregionalism (Thayer 2003; Snyder 1974). Deep Ecology was a philosophical movement which advocated the intrinsic value of places—a human ecology that was *not* human-centered. It promoted a reverence for life ethic which included all carbon-based organisms.

The Gaia hypothesis, noted above, was named after the ancient Greek Earth goddess, and portrayed the biosphere as an integrated, self-regulating entity. If imagined this way, the prevailing mode of production and development compromised the planetary system, and would eventually die out. The "Mother Earth" trope was frequently invoked as evidence of the sacredness of place. This was true, for example, among Indigenous movements across the globe who adopted this perspective to establish their rights to human dignity, and to extoll the sacredness of the land (Gill 1987).

Bioregionalism was for all intents and purposes the default political philosophy of dark green religious actors (Taylor 2000, 2002). Advocating an economy that worked *in place*, or rather in *places*, bioregionalism reduced spheres of economic and governmental activity to areas defined by ecosystem boundaries, rather than arbitrary political maps. The Pulitzer Prize winning author Gary Snyder was one of its main proponents, and he helped to engineer international conferences devoted to realizing the vision of networked bioregions.

In what might have been the opposite of that vision of a living planetary system, James Watt, the Secretary of the Interior from 1981 to 1983, vociferously argued that the environment was a footnote in the history of humankind. Nature's destruction was required for the establishment of God's new kingdom on a new Earth. Watt carried his Christian baggage with him into his job, riding the well-worn claim that environmental religions were pagan and pernicious, and that they should be eradicated. Radical environmentalists used Watt as a proxy for all anti-environmental Christians. The execution of his duties, according to Watt, required "follow[ing] the Scriptures, which call upon us to occupy the land until Jesus returns" (quoted in Glass 2005: 1043–1044).

It is perhaps more difficult to characterize countercultural movements than international religious and political declarations, simply because of the diversity of perspectives and practices enacted to foster resilient and sustainable societies. The review above merely offers a superficial treatment of those areas where environmental religions thrive. The voices of the scientists and climate alarmists have become more urgent over time, even as anti-environmental responses grow increasingly strident. Ultimately, the future lies not in the hands of those in the ivory tower, but in the hands of readers who will hopefully grow to be informed citizens of a multi-perspectival planet.

Outline of the Chapters

Edward Burnet Tylor (1832–1917), featured in Chapter 1, was widely considered one of the most impactful anthropologists, even if he was not a very good anthropologist. His primary sources were the accounts of missionaries and explorers, which he weaved into explanations of the curious cultural production occurring among the ethnic, cultural, and ethical others that were the targets of proselytization and "discovery." For Tylor, a key to those cultural complexes was Animism, which he imagined as the mistaken attribution of agency to inanimate things—for example, the idea that features of the landscape might have spirits. Animism was a product of what Tylor called "soul coughing," experiencing the "self" as somehow disembodied. But for Tylor, the existence of a soul was a nonempirical claim, people did not actually have a soul, which led him to the conclusion that it was a cognitive error. Because of his focus on mistaken cognition as the foundation of religion, he might be classified as an idealist in orientation, placing greater explanatory emphasis on thoughts and ideas rather than practices and politics. Tylor believed that such primitive religions would eventually be eclipsed by polytheistic, and then monotheistic religions (considered more rational by Western Europeans than polytheism).

The sociologist Emile Durkheim (1858–1917), the star of Chapter 2, was one of the most pivotal characters for those who trace the theoretical contours of religion and nature for several reasons. He posited that all societies made a distinction between the sacred and the profane and that these were originally mediated by natural symbols or animal totems. In contrast to Tylor's notion that religions would evolve toward more rational explanations, Durkheim classified societies into those characterized by mechanical solidarity, and those which exhibited organic solidarity. Those social modes might provide a starting point for contemporary critiques leveled at environmentally destructive enterprises. Finally, Durkheim believed that religion was the womb of our categories for the world, and the firstborn category was that cognitive distinction between sacred and profane that made society possible. Durkheim held that there were social facts that were basic elements of society that were unavoidably true and unescapable. While Tylor speculated that religions had their origins in a mistaken projection of agency, Durkheim claimed that the original religion was totemism, the association of natural elements (typically biological) with the sacred. Taboos helped ensure their purity. Social formations also had an important geographic element—they were social constructions specific to the habitats in which they took shape (Durkheim 1965 [1915]). This is important to understanding Eliade's arguments in the following chapter.

In Chapter 3 we encounter the scholar of religions Mircea Eliade (1907–1986), one of the most influential religion scholars of the twentieth century. His comparative work fundamentally depended upon Durkheim's distinction between the sacred and the profane. Where Durkheim had failed by reducing religion to its social function, Eliade said that religion was *sui generis*, in its own category, and might only be properly approached and understood by those who had specialized knowledge in this area. Importantly for scholars of religion and nature, it was Eliade who famously declared that the sacred would irrupt into the profane world, making obvious the distinction between them. This created an *axis mundi*, or world center, for those who recognized the event as a sacred intervention into the profane world (Eliade 1959). Drawing on the theologian Rudolph Otto (1925) (who really took his cues from earlier work by the philologist Max Müller), Eliade declared that religion derived from an experience in or with nature that was sublime, terrifying, awe-inspiring, or otherwise emotionally overwhelming. Interestingly, some Indigenous and other epistemologies that vie for political recognition in land sovereignty claims continue to use derivatives of Eliade's ideas about sacred space to make legal cases related to land.

The scholar Rosemary Radford Ruether, highlighted in Chapter 4, worked from within a particular religious tradition, noting how the patriarchal and oppressive nature of Christian theology resulted in a parallel exploitation and oppression of women and natural systems. Much of Ruether's early theological work focused on communities of the oppressed (Ruether 1972, 1988, 1996). She combined her interests in liberation theology with a theology of resistance to the exercise of secular power (Ruether 1998, 2005, 2007). Ruether excavated histories of the Abrahamic religions and noted the persistence of male-centeredness and human supremacy over the environment within its historical unfolding (Ruether & Ellis 1990, 2007). Others followed Ruther, pushing for strong reform of Christian traditions.

Chapter 5 is dedicated to Vine Deloria (1933–2005), who was born on the Standing Rock Sioux Reservation (Lakota) not too far away from the Black Hills. He was one crucially important vector for the dissemination of American Indian scholarship in the United States in the late 20th and early 21st century. Deloria was a central figure in exposing the American public to American Indian religions and lifeways. He articulated the legal obstacles and pitfalls encountered when Natives spoke about their religions to an ostensibly secular government. Deloria also called attention to the public land disputes between American Indians and extractive industries. His work analyzed historical claims of sovereignty and the uneven application of laws related to religious freedom. He provided insights into how Indigenous knowledge might help to shape policy recommendations going forward, especially in tribal territories. Deloria was situated at a pivotal time and place for American politics, Native sovereignty movements, and direct political action.

In Chapter 6 the focus is on Carolyn Merchant (b. 1936), who was sympathetic to Indigenous resistance to the impositions of empire. She was a powerful force in environmental history, putting together sweeping analyses of environmental ethical ideas. She also offered prescriptions for how to escape what she perceived as interlocking forms of oppression: the human domination of earth systems, and the systematic devaluation of women, which might only be corrected by holding up women's ecological knowledge and practice. Her well-known *The Death of Nature: Women, Ecology, and the Scientific Revolution* (1980) exposed the androcentric (male-centered) and often violent imaginaries related to nature. She referred to philosophers like Francis Bacon, who explained the scientific enterprise as a struggle against a feminized "nature," where natural facts must be, quite literally, ripped from Nature's breast. Some authoritative histories remain important contributions to the religion and nature canon. Clarence Glacken's (1967) important work on antiquity to the present, and Roderick Nash's *Wilderness and the American Mind* (1973) are mainstays in this regard. But those who drank deeply from Merchant's well of wisdom also encountered feminist, postcolonial, and multi-cultural histories of science and ethics which continue to shape the field of religion and nature today.

One of the grassroots movements Merchant wrote about was the Kenyan Greenbelt Movement. Wangari Maathai, the subject of Chapter 7, was an advocate for the role of women in maintaining land sovereignty and ecological restoration. She directed a handful of nonprofit agencies in Kenya which facilitated the planting of tens of thousands of trees in deforested areas where there was little arable land and few resources from which residents might scrape a living. The most well-known of these organizations was the Green Belt Movement, which later expanded beyond Kenya to other African nations. Tree planting simultaneously addressed several problems for the people. First, the trees grasped the soil, thus preventing erosion. Second, they provided a source of fuel wood, which was increasingly scarce. Along with planting trees, Maathai and her followers also diversified the crops which were grown in those places, and the animals that utilized the trees as habitats. Maathai was important for implementing a successful ecological movement within one of the "world religions," Christianity. But Maathai's own writing wrestled with the imposition of Christian rationality against the backdrop of entrenched commitments to traditionalism. In addition, her work ran up against the patriarchal culture of her homeland. She experienced harassment from civic and religious authorities because, as a woman, she was disregarding the "natural" order between God, men, and women. Her work and legacy led to a Nobel Peace Prize, and her movement became a spark for women-led restoration movements beyond Kenya. Little attention focused, however, on the relationship between institutionalized religion and local traditionalist practices, which sometimes conflict, and at best may be uncomfortable bedfellows.

In his posthumously published *A Sand County Almanac* (1949), the ecologist Aldo Leopold shared a story that had shifted his whole perspective—the tale of an emotional eye-to-eye encounter with a non-human animal. He had shot but not killed a she-wolf who was with pups and arrived beside her broken body in time to "watch a fierce green fire die" in her eyes. The first dissertation written about the emergent field of environmental ethics was J. Baird Callicott's elaboration of Leopold's ideas and their application in the conservation in the United States. Callicott (b. 1941), the thinker in Chapter 8, was a player in a long-running internecine dispute among environmental philosophers regarding whether it was possible to generate an internally consistent non-anthropocentric ethic—that is, an ethic that was *not* human-centered. He advanced a sophisticated ecocentric (ecosystem-centered) ethic. Callicott also engaged in a comparative projects, examining environmental ethics from various cultural perspectives (Callicott & Nelson 1994), and wrote for other edited volumes that did the same (Callicott & Nelson 1998, 2004). The lesson he offered was that some glimmers of hope from other cultures and their religious complexes might help us learn how to address the core problem of environmental ethics: that conservation was getting nowhere because it clung to Abrahamic concepts of human superiority and control.

The Indian physicist-turned-activist Vandana Shiva connected the dots between food, the capitalist mode of production, and marginalized populations, and her theoretical contributions are the subject of Chapter 9. While

India was her home, Shiva's work illuminated places all over the world where similar contestations occurred. Specifically, Shiva provided a voice for the poor, especially for poor women worldwide who, in their poverty suffered the additional injustice of being born into a patriarchal world. It was a care for food systems and the rural poor which spurred her contributions to another academic discourse, ecofeminism. The primary targets of her critique included modern capitalism and development schemes that reduced biodiversity and further disenfranchised people who were already scraping a living from resource-scarce landscapes. She also attacked the increasing use of genetically modified foodstuffs, highlighting the problems with agriculture that required massive inputs of fertilizer and pesticides. Shiva's own farm, Navdanya, was a woman-led resurrection of an agricultural practice adapted to place and grounded in traditional spiritual principles.

Jane Goodall (b. 1934) rose to prominence for her work on chimpanzee conservation and behavior, and she is the inspiration for Chapter 10. She is at this writing a United Nations Goodwill Ambassador, author, and a global advocate for the recognition of morality and spirituality among our closest primate ancestors. Her observations of tool use passed down culturally within chimpanzee social groups erased the long-popular notion that tool-making was what made humans unique. Much of her well-publicized work with chimps was dismissed as insufficiently rigorous by some of her colleagues, for she lived among her subjects, and knew them as individuals who demonstrated empathy, compassion, and cooperation. Fighting then against both critics in her own field and against widely held perceptions of human superiority, she advanced her own global-scale spirituality of hope. Goodall and others contributed to the increasingly obvious facts: (a) that social groups (including primates) can have morality without religions; and (b) given animals' modes of empathetic and emotive relations within and between their social groups they exhibit spirituality, or at least proto-spirituality. In the academy, the Religion and Animals Consultation (and later Group) was launched as one of the subgroups of the American Academy of Religion (AAR). The "buzzing, breathing, clicking, clacking, biting, stinging, chirping, howling landscape of religious studies" (van Horn 2011: 230) has diversified in the subsequent decades as explorations from the biblical to the behavioral, from the aesthetic to the cognitive have blossomed.

These ten thinkers are pivotal to the emergence of scholarship at the intersection of cultures, religious social formations, and nature (however conceived). The concepts pioneered by them are parts of active and ongoing research programs in the present. Each case provides a different disciplinary lens, and each represents an attempt to theorize the relationships between humans, and between humans and their habitats. In what follows, the reader will encounter the specifics of their approaches, and problems with or critiques of them. Through the thinkers collected here, students will grasp the development of this academic field and the theories and methods employed therein.

Notes

1 The use of the term "environment" rather than, say, ecology or religion, might be criticized by pointing out that its use implies some boundary between humans and "the environment." Many key thinkers have pointed to this sort of dualism as the root of ecological crises. I am aware of such criticisms but delving into the history of debates about the use of such terms is beyond the scope of this book. Students might just recognize that each of these terms has its advantages and drawbacks.

2 The Baha'i tradition emerged in the early 19th century and was grounded in the unity of the divine, and the inherent worth and dignity of all humans. By their reasoning, the great religious figures of history (Jesus, Mohammed, etc.) all pointed toward the same God, they were simply different cultural manifestations of that divinity. For this they were widely persecuted.

3 See the Council on Foreign Relations' (CFR) Global Conflict Tracker, https://www.cfr.org/global-conflict-tracker, accessed 17 July 2023.

4 These are unpacked in Chapter 8. Most humans have anthropocentric—human-centered—ethical sensibilities. This contrasts with bio- or eco-centric models, where landscapes or ecosystems are imagined as morally relevant.

5 The *Anthropocene* was a term first used by the economist Will Steffen to designate an epoch in Earth history where all Earth systems were dominated by human activity. The previous age was the Holocene (approximately 11,600 years ago to the modern era).

6 There are significant overlaps with the economist Thomas Malthus who wrote around the turn of the 19th century. In fact, Hardin and others were often tagged with the label "neo-Malthusian." Most of those in the modern era who made such arguments left out Malthus's obsession with using population control as a strategy for eliminating the poor and those without utility.

7 Agenda 21 was targeted by anti-globalization advocates who perceived in it a creeping form of global governance.

8 See https://fore.yale.edu/FaithActionSDG, accessed 7 August 2023. The Forum on Religion and Ecology (FORE) has the most extensive list of the global religions (world religions) and their specific activities.

9 See https://fore.yale.edu/Resources/Multimedia/Video/Earth-Charter, accessed 7 August 2023.

1

EDWARD B. TYLOR

"From Human Phantoms to Animal Persons" (Anthropology)

The Original Religion: Animism

Edward Burnet Tylor (1832–1917) is widely considered one of the most impactful anthropologists, even if he shied away from fieldwork. He was raised in England as a Quaker and much of his youth was consumed by illnesses. In his convalescence-focused travels, he befriended another Quaker who was able to convince the young Tylor to accompany him to Mexico for his archaeological work on Olmec and related civilizations. Tylor was so impacted by the experience that his recounting of that expedition was recorded in his first published work, *Anahuac, or, Mexico and Mexicans, Ancient and Modern* (1994 [1861]). Following those travels, his primary sources were the accounts of missionaries and explorers, which he weaved into explanations of the curious cultural production occurring among the cultural, ethnic, and ethical "others" that were the targets of proselytization and discovery. He proposed a "science of culture" which would investigate the "knowledge, belief, art, morals, law, custom, and any other capabilities and habits acquired by man as a member of society" (Tylor 1883: 1). Tylor was expressly interested in differences between the "civilization of the lower tribes as related to the civilization of the higher nations" (Tylor 1883: 1). Like most of the early social scientists, Tylor was influenced by evolutionary theory, and he viewed it as a complementary enterprise. In his own work, Charles Darwin cited Tylor—they were both engaged in the popularization of evolution.

The origin of religion, Tylor thought, was the experience of the "self" as somehow disembodied yet efficacious—that is, the disembodied self could act in the physical world. Imagine, for example, a dream in which you are the central figure, but you see yourself acting in the dream in the third person.

DOI: 10.4324/9781003364801-2

Or perhaps you have a perception that you have some essence, some soul that remains even if the terrestrial body withers. This, Tylor thought, was a universal human trait, so that something like an experience of this "soul" was the origin of religion. According to Tylor, the primitive tribal peoples of the world were arrested in this less evolved mode of religion, where they perceived that other creatures and features of the landscape similarly had spirit selves. Tylor himself had no belief in a soul but imagined this was the form that primitive religions took.

Tylor was in part responding to some earlier scholars who had begun to search for the origins of, and the comparison between religions, especially F. Max Müller (1823–1900), a philologist (linguist) who 12 years earlier proposed a "science of religion" *(Reigionswissenschaft*, in German) to discern religion's origins in humanity, and its natural course. Müller was transparent about the aim of this project: "it will show for the first time fully what was meant by the fullness of time; it will restore to the whole history of the world, in its unconscious progress towards Christianity, its true and sacred character" (Müller 2002: 74). Although Christianity, Müller believed, was the most rational and evolved religion, all religions had a common experiential root: "a groaning of the spirit, a struggle to conceive the inconceivable, to utter the unutterable, a longing after the Infinite, a love of God" (Müller 2002: 175–176.) For Müller, religion was an ultimately untranslatable experience triggered by a subjective human faculty that experienced the perception of the infinite in nature. Such encounters with the infinite could excite as well as terrify, but they were always aroused by experiences that defied our linguistic capacity to capture them.

Müller's penchant for Christianity was apparent in his publications, and for Tylor it was one of the distasteful turns that characterized too many scholars seeking after religions' origins. Both Müller and Tylor assumed the evolutionary model that was spreading through the academy, but clearly, it could be interpreted in very different ways. Müller imagined that the evolution of religion and society pushed inexorably toward Christianity and that other more primitive civilizations would move from simple and magical to rational and monotheistic. On the other hand, Tylor took for granted the notion that the "lower civilizations" would move toward higher modes of religious sentiment but rejected Müller's notion that Christianity or any other religion would emerge preeminent. Tylor believed that in "higher civilizations" humans came to trust in science as more ostensibly divine action was explained away by science. Subjected to Tylor's science of culture, Christianity received no favored status but was rather analyzed with the same tools deployed to understand lower civilizations.

But like Müller and others of his day, Tylor was excited by the search for the origin of religions, which he attributed to "a doctrine of souls and other spiritual beings in general"—Animism (Tylor 1883: 23). He worried that most definitions of the category religion would unduly exclude some cultures by prioritizing some element of the development of religions, "not the deeper motive which underlies them" (Tylor 1883: 424). The best way to proceed with a science of religion was a simple operationalization of "religion"—which led Tylor "simply to claim, as a minimum definition of Religion, the belief in Spiritual Beings" (Tylor 1883: 425). Animism has one doctrine: "Animism, in its full development, includes the belief in souls and in a future state [after death] ...practically resulting in some kind of active worship" (Tylor 1883: 427). Our everyday experience of the difference between our waking lives and dreams about ourselves, where our "phantom" (Tylor's term) appears, the notion that other agents within the environment also have phantoms, and that there is a system of exchange between them, is enough to make Animism a plausible explanation. We lead a double life—one mundane existence where we shop, eat, sleep, and poop—and another where our souls (phantom bodies) cavort and compete with those of other spirits and animals. If many animals, places, and people were animated by a soul whose agency extends to both the spiritual and material world, that is Animism. But remember, Tylor wanted his approach to culture to be scientific, which led him to the conclusion that such attributions of agency to others were erroneous, a mistaken perception.

As Tylor put it, "at the lowest levels of culture of which we have clear knowledge, the notion of a ghost-soul animating man while in the body, and appearing in a dream and vision out of the body, is found deeply ingrained" (Tylor 1883: 499). This is a sort of biological categorical imperative—something irreducible to other phenomena: "The animism of savages stands for and by itself; it explains its own origins" (Tylor 1883: 500).[1] Note this language. To give a glimpse of what's to come, many other scholars who claim to have found the origins of religions rely on some foundational element that is *sui generis*, something so unique that it inhabits its own category. Whether it was the individual psyche, the social group, the biological kin group, or some other irreducible thing that did not require its own explanation, it was a familiar way in which scholars attempted to understand religious cultural production.

Unlike those who viewed the primary unit of analysis as the social collective, Tylor was concerned with individual persons. For Tylor, the seed of religion is planted in the individual psyche. Emile Durkheim (Chapter 2), for example, was a methodological holist, one who took the social collective, rather than the individual, as the unit of analysis. For scholars of Religion and Nature, three elements of Tylor's thought are particularly relevant: (1) he popularized the term *animism* as a term for the original religion; (2) he suggested that culture is disseminated through individuals' effects on other individuals; and (3) argued that science has explanatory power over myth and text.

The anthropologist James G. Frazer was influenced by Tylor's writings, particularly his idea that religions would ultimately be eclipsed by science. His masterwork, *The Golden Bough* (1890), restated some of those ideas forcefully. Where Tylor believed animism was the original religion, Frazer referred to the "worship of nature" as the first manifestation of religion (Frazer 2012 [1926]: 17). Like Tylor, he believed that so-called primitive religions were "anthropomorphic superstitions" (Taylor 2005a: xiii) which would be eclipsed by polytheistic and monotheistic religious complexes. The ultimate end, though, was a "despiritualization of the universe" as Western cultures spread across the globe (Frazer 2012 [1926]: 9; Taylor 2005a: xiii). For those savages' religion had a strong experiential and emotional component which was being gradually erased by a bureaucratized and increasingly rational and secular world. These were really the first stirrings of the secularization hypothesis, which imagined the industrialized world growing increasingly secular over time.

Frazer also introduced *sympathetic magic* as an important element in the emergence of religion. Sympathetic magic assumed some contact between agents, not all of whom were human. The magician, the holy practitioner, was a proto-scientist, engaging in rudimentary but nonetheless empirical analysis of the rest of the natural world. Frazer also postulated a form of contagious magic, where the elements were proximate and laws were associational, and a homeopathic magic based on metaphorical and metaphysical reasoning. For instance, contagious magic is expressed in rituals which imagine that things like the umbilical cord, placenta, and afterbirth are believed to remain in relation with a person throughout their lives. On the other hand, homeopathic magic suggests that items that resemble, or have contacted a body, can have effects on it.

The importance of the category magic pervaded the anthropological and religious studies literature for the next century taking different forms, including holy spirits/forces (such as *mana*, *manitou*, and *wakan*) and material culture (totems, fetishes, amulets, and more) (i.e., B. Malinowski 2004, 2020). Each of these preternatural forces was imbued with irreducible agency and was thus beyond further investigation. But what we ultimately found in each of these manifestations of magical powers is the underlying exercise of mundane power, social, economic, and political contestations hiding under the mask of magic (Styers 2004; Harvey 2013).

Tylor's work was also influential on early psychologists. His insights regarding the centrality of the individual psyche in the emergence of religion directly influenced Sigmund Freud (1964, 1998) and Carl Jung (1964) to a lesser extent. Freud was, like Tylor, concerned with individuals and believed that the root of religion was a mistaken projection of agency. For Freud, the mechanism for this was a biological trait, and its trigger was internal to the person (as with Tylor). But for Freud, religion was basically a neurosis and could therefore be treated. Thus, in contrast to Tylor, Freud was a materialist—there was no spirit realm to which religion might be abstracted. Further, for Freud, the future of this illusion—religion—was extinction, thanks to refinements of the scientific method (Freud 1930; 1964 [1927]). The primordial man (the gendered noun follows Freud's use) invented religion as an analog for neuroses, to explain forces of nature that were mystifying or dangerous. The civilized human was bent on subduing and exploiting nature. The natural state of humankind in Freud's view was to either be in terror of the natural world or intent on overcoming it. These remained important ideas throughout the twentieth century (York 2005: 673).

It is possible to see elements of Tylor's approach in each of these with small nuances. Ultimately though, Darwin's influence on Tylor is evident in the supposition that attribution of agency to nonliving things should be imagined as quite natural, even if mistaken. As Darwin put it:

> The tendency in savages to imagine that natural objects and agencies are animated by spiritual or living essences, is perhaps illustrated by a little fact which I once noticed: my dog, a full-grown and very sensible animal, was lying on the lawn during a hot and still day; but at a little distance a slight breeze occasionally moved an open parasol, which would have been wholly disregarded by the dog, had any one stood near it. As it was, every time that the parasol slightly moved, the dog growled fiercely and barked. He must, I think, have reasoned to himself in a rapid and unconscious manner that movement without any apparent cause indicated the presence of some strange living agent, and no stranger had a right to be on his territory.
>
> *(1871: 67; quoted in Bulbulia 2004: 3)*

Religion does not get much attention in Darwin's *The Descent of Man* (1871). But when it does come up, Darwin assumes it is merely a peccadillo expression of the evolved brain that had no real effect on natural selection. Perhaps it becomes clearer why early social scientists taken with Darwin's theory might imagine a social evolution from primordial religions grounded in reverence for places or natural objects, to their future extinction in explanatory sciences.

Related Research Programs: Cognitive Science of Religion and Evolutionary Psychology

The notion of animism and Tylor's idiosyncratic version of it was chastened, if you will, by contemporary scholars. There are two streams of scholarship that emerged in the late twentieth century. More than a century after Tylor, the scholar of religions Stewart Guthrie's *Faces in the Clouds: A New Theory of Religion* (1995) reinvigorated the notion that religion was an attribution of agency to things that were not typically agential. Guthrie called this systematic anthropomorphism. Anthropomorphism is viewing non-human persons, places, or things, in human ways, for example, by saying that we might perceive shapes and faces in the clouds, even though clouds are not really "persons."

Guthrie put forth three ways in which scholars of religions might focus on humanistic theories of religion:

1 The Wish-Fulfillment Group, which posits that religion alleviates emotion and anxiety. Examples of this might be Freud or other psychologists.
2 The Social Functionalist Group, whose argument was that religion provided some sort of social solidarity. Into this category, we might put the sociologist Emile Durkheim, or the anthropologist Claude Levy-Strauss. From this view, religion served the *function* of facilitating and preserving social relationships and interdependence.
3 The Intellectualist Stream, which suggested that religion is "an attempt to interpret and influence the world, a task it shares with science and common sense" (Guthrie 1995: 10).

Guthrie planted himself firmly within the third camp and began his book with a nod to Tylor's popularization of the term animism, even if he distanced himself from Tylor's understanding of the term. He argued that the sort of systematic anthropomorphism he described was preconscious, and therefore not a projection of the gods in the image of the human. Rather it was an evolved propensity to see agents where there were none. Our supremely social primate species easily sees faces where they are not (the man in the moon, faces in the clouds, etc.) because agents who might communicate, help us, or do us harm are important to our survival. If we mistake the boulder for a bear and flee, what have we lost? A handful of calories. If we make the opposite mistake, and take the bear for a boulder, we are at risk of losing a lot more than a few calories. But it is this preconscious element that led Guthrie to the conclusion that systematic anthropomorphism was natural, and that it might explain other elements of human cultures. As Guthrie put it:

> Seeing that perception is interpretive also enables us to see anthropomorphism not merely as an idiom for conscious explanation, but as the result of a general, spontaneous, and unconscious interpretive tendency. This tendency, general among animals, is to find as much organization as possible in things and events.
>
> *(Guthrie 1995: 37)*

Cognitive scientists and evolutionary psychologists took this claim seriously, adding both experimental and ethnographic evidence that anthropomorphism was related to two fundamental elements of what they believed to be a modular mind. The first was the Hyperactive Agent Detection Device (HADD), a notion that Guthrie advanced, but was later elaborated experimentally (Barrett 2004). But the psychologist Pascal Boyer surmised that there must be other innate systems involved in the construction of spiritual beings. Specifically, the HADD had to be complimented by a Theory of Mind (ToM). Because humans only know what it is like to be a human (and not a bat, or a cat, or a bear), we infer that those other over-detected agents in our midst have a mind roughly like ours, one that is goal-oriented or purposive. Further, Boyer demonstrated that what he called counterontological or minimally counterintuitive concepts were crucial for the transmission of religious ideas. Concepts that violate one or two, but not several of our ontological categories, like a person who can walk through walls or cats who surf, are memorable and are more easily transmitted.

Boyer concluded that religion must be a "spandrel," something parasitic upon other naturally evolved cognitive proclivities (such as the HADD and ToM) (Guthrie 1995; Boyer 2001; Atran 2002).[2] For him, religion derived from a set of natural cognitive responses to environmental stimuli that were free riders on functional, and thus adaptive cognitive and neurophysiological architecture. In their explanatory models, functional features of human cognitive processing (i.e., the HADD hypothesized by Barrett and others) predisposed humans to "see" agency in inanimate objects. Extending such agency detection to invisible or otherwise preternatural beings represented a natural response from a developmental perspective, but one that was a by-product of the evolved tendency to detect agents in a variety of natural objects. Importantly all of those working in the cognitive science of religion and evolutionary psychology imagined that the roots of religion were biological (i.e., they were hard-wired in our primate brains). None of those scholars dismissed the cultural pressures that give rise to variations in their expression, but religions are not epiphenomenal cultural impositions upon individuals.[3] They are colorful decorations which depend upon leveraging other adaptive features of our evolved brains.

Others have imagined that religion might indeed be adaptive. For instance, the anthropologist Michael Winkelman argued that shamanism was the root of both religion and ritual healing (Winkelman & Baker 2010;Winkleman 2009). The term shamanism itself has been the subject of much scholarly debate, for it has been widely applied outside the geographical range in which its practice was first marked. In the earliest anthropological literature referring to shamanism, the religious complexes were examined in modern-day Siberia (Dioszegi & Hoppal 1995, 1996, 1998). However, the term has been used to refer to a wide variety of cultures in multiple locales. Winkelman argued that it is a term that has cross-cultural validity because it referred to a set of practices common to forager cultures across the globe (Winkelman 2009). Shamans are practitioners with special knowledge of how to heal individuals and even social groups by maintaining right relations with the rest of the non-human biotic community. The neuroscientist Daniel Levin and others argued that highly emotional events such as ritualized performances may activate opioid systems, fostering a sense of attachment and fellow feeling (Mithen 2007: 216; Levitin 2008: 198). These experiences may catalyze integrative modes of consciousness characterized by brain discharges which cross the neuroaxis and inhibit some of the brain's regulatory mechanisms. There is heightened communication between the attentional, behavioral, and emotional centers of the brain, synchronizing their discharges and projecting these neurological firings into the frontal lobes (Winkelman 2009: 472). Right relations are promoted through ritual practices which depend on drumming and song, stimulating cross-hemispheric connections in the brain and an altered state of consciousness (ASC). The integrative mode of consciousness is crucial to fostering relationships between humans and between groups of humans and the habitat.

Other approaches for how religion might be biologically adaptive include investigations into the emergence of long-duration running in early *Homo* species (Sands and Sands 2009), where it was said that the practice of persistence hunting produced an ASC which facilitated highly affective and reflective relationships to landscapes. The psychologist Joseph Bulbulia, the sociologist Richard Sosis, and others have contributed significant research which illustrates the important role that costly signaling plays in religious transmission (Sosis 2003; Sosis & Alcorta 2003; Bulbulia 2004: 64–72). Costly signaling is the practice of performing acts that are dangerous to the individual, but which illustrate their commitment to the group. For instance, walking across hot coals is an important ritual for some peoples. The Lakota Sun Dance, where dancers have their flesh cut off, and some participants pierce their pectoral muscles with eagle claws, is another example. They provide evidence of co-evolved biologically entrenched behaviors that can facilitate the development of more resilient ecosystems (Kiper & Sosis 2009).

The sociologist David Sloan Wilson (2002) proposed a different way of understanding religions, by arguing that they promoted higher levels of prosociality than those without religious cohesion. When their ideas fostered moral and intellectual cohesion, they enacted a collective vision that allowed them to out-compete other social groups and populations. Although there remains some disagreement between those who view religions as adaptive and those who believe they are mere by-products of natural human cognition (Boyer 2001; Atran 2002; Bulbulia 2004), ideas rooted in Tylor's speculations grew into testable theories of evolutionary psychology which have been influential, if contested, areas of fruitful study.

Related Research Programs: The New Animism and the Religious Lives of Other Animals (and Plants)

A second stream of research emerged that was also to some extent indebted to Tylor. He of course popularized the term animism. But those pursuing what was, by the twenty-first century, being called the new animism wanted to preserve the name without the theoretical baggage from the nineteenth century. The scholar of religions Graham Harvey was instrumental in fostering the growth of that area of scholarly inquiry in the Academy in Europe and in the United States. He published well-received books including *Animism: Respecting the Living World* (2006), the *Handbook of Animism* (2015), and *Food, Sex, and Strangers* (2013), all of which were important contributions to the intellectual outgrowth of new animism. Earlier contributions that were crucially important include Philippe Descola's *Beyond Nature and Culture* (2013, first published in French in 2005). Descola introduced readers to the problems that arise when other-than-human animals are considered to be "persons" just like you and me, except in a different form. They share with us a culture and a set of customs for maintaining right relationships in their habitats. Different ontologies and epistemologies must first be understood before their practice can be explained. Harvey, in an appreciative application of the anthropologist Irving Hallowell's ethnology, noted which animals, plants, and even rocks were "alive" required some prior experience of them—they were socially salient and agential. The perception of the boundaries between human and other-than-human differs.

The anthropologist Vivieros de Castro's work challenged previous research not by focusing on the separation of nature and culture, but rather on each specific agent's own viewpoint—what he called "perspectivism." There are monkey persons and human persons, and they share a similar culture even if they inhabit different forms. The monkeys (and birds, and snakes, etc.) also have a shaman in their midst whose job it is to consort with human person shamans to ensure balance. If a hunter takes more than they need, or if they are careless or not appropriately appreciative of the animal for giving itself to the hunter, those other animal persons will withdraw their support from the human persons, or seek revenge on them through accidents, poisonings, and the like. Perspectivism was influential on anthropology, but the anthropologist Robin Wright pointed out some of the limits of perspectivism (Wright 2014). Wright contended that perspectivism does not explain some of the ritual practices of some Amazonian cultures, nuancing the theoretical landscape.

Within the academy, the Animals and Religion Consultation (which later became a "group") was introduced in 1993. The unit has since fostered a wide range of perspectives on the relationships between religions and animals, including Indigenous traditions, mainstream global religions, and some new religious movements. Some of this academic production utilized ancient religious texts to reimagine how other-than-human animals were important to the religious lives of humans historically. Others used endangered or other charismatic species as mascots to activate environmental concern. In other cases, landscapes were conserved because protected animals were living there. Finally, some scholars have developed rituals specifically designed to challenge human-centered perspectives. For instance, the Council of All Beings, pioneered by activists and authors Joanna Macy and John Seed, was a ceremony in which participants acted as "the voice" of a particular animal, raising that animal's survival needs and concerns with the group. The point was to de-center humans and to include, even if symbolically, other animals.

These different research programs are in some way related to, or derivative of Tylor's early, and problematic anthropology. Fortunately, some contemporary scholars are resuscitating animism, and developing rich avenues for further research.

QUESTIONS FOR DISCUSSION:

- What are the key differences between Tylor's definition of animism, and later uses of the term?
- Guthrie's contention that religion is anthropomorphism was an argument from a scholar from the humanities which was adopted and tested by scientists. Can you think of other examples either (a) where scholarship in the humanities advanced scientific research, or (b) where scientific ideas were effectively translated into richer, more evocative media or prose?

Notes

1 The categorical imperative is a phrase utilized by the philosopher Immanuel Kant (1724–1804) to describe fundamental rules or laws. For Kant, all moral reasoning was ultimately justified with reference to these imperatives. In suggesting that Tylor made these "biological" I am suggesting that what was for Kant an acceptance of the limits of human reasoning was, for Tylor, something innate in all humans.

2 A spandrel is an architectural term which refers to the space between the supports of two adjoining arches. The space serves no real architectural function but was often decorated with ornate designs, sculptures, or paintings. The biologist Stephen Jay Gould used this term to describe religion, which is not necessarily functional in the evolutionary sense, but which, when "decorated" with cultural accompaniments, draws more attention than do the underlying functional traits that support it. Darwin's dog enforcing his territorial boundaries against an offending umbrella was perhaps the first exposition of a spandrelist explanation for human religiosity (1871 [1981]: 64–66). In his discussion, he cited theories of religion offered by E.B. Tylor and Herbert Spencer.

3 If culture is not epiphenomenal, it simply means that there is not some *thing* that is culture that is above, or beyond the embodied humans who manufacture cultural forms.

2

EMILE DURKHEIM

"Religion Is the Social Glue That Holds Society Together" (Sociology)

Religion and Society: Neither Is Possible Without the Other

The sociologist Emile Durkheim (1858–1917) is one of the most pivotal characters for those who trace the theoretical contours of religion and nature for several reasons, including his distinction between the sacred and the profane, and the classification of societies into those characterized by mechanical solidarity, and those which exhibited organic solidarity. Standing at the roots of both structuralism and functionalism in the social sciences, Durkheim set out to explain that the origins of religions and societies were entangled and that each made the other possible. Religion was the womb of our categories for the world, and the first-born category was the cognitive distinction between sacred and profane that made society possible.

When the wheels of industry began to turn, they also began to creak as the specialization of labor and the compression of social relations in time and space raised the prospect of social collapse. In the industrial age, a new mode of social arrangement became prevalent globally, and religion was the grease that tamed the noise coming from the manufacture of social solidarity. Durkheim defined religion as:

> a unified system of beliefs and practices relative to sacred things, that is to say, things set apart and forbidden—beliefs and practices which unite into one single moral community called a Church, all to adhere to them.
>
> *(Durkheim 1965 [1915]: 62)*

DOI: 10.4324/9781003364801-3

To begin to explain Durkheim's primary scholarly preoccupations, it is useful to begin with the fundamental distinction between the sacred and the profane. According to Durkheim, "the fundamental categories of thought, and consequently of science, are of religious origin" (Durkheim 1915: 466). His core epistemology should be imagined as a sort of "social Kantianism."

There is *a-priori* knowledge, but while the philosopher Immanuel Kant imagined that *a-priori* truth dwelled inside the structure of the individual brain, Durkheim imagined that it came from society. As he put it, (1915):

> the division of things into sacred and profane, the notions of the soul, of spirits, mythical personalities, and of a national and even international divinity, a negative cult with ascetic practices which are its exaggerated form, rites of oblation and communion, imitative rites, commemorative rites and expiatory rites; nothing essential is lacking.
>
> *(Durkheim 1915 [1965]: 462)*

If, as we may accept, the synthesis *sui generis* which every society constitutes yields new phenomena differing from the those which take place in the individual consciousness," Durkheim tells us, "we must also admit that these facts reside exclusively in the very society itself which produces them, and not in its parts—that is, its members.

(Durkheim 1972: 69)

> There is nothing below or behind the social that we can reduce the study of religion to, for religion is the *conscience collective* of the society itself—its perception and reproduction of its ideal self.
>
> *(Durkheim 1965 [1915]: 470)*

It is helpful to dig deeper into the "facts" Durkheim highlighted—*social facts*. Social facts were basic elements of society that were unavoidably true and inescapable. While Tylor speculated that religions had their origins in a mistaken projection of agency (animism), Durkheim claimed that the original religion was *totemism*, the association of natural elements (whether geographical or biological) with the sacred, and taboos helped ensure their purity. Into the totem, the irreducible sense of the collective self was projected. But "social facts do not differ from psychological facts in quality only: *they have a different substratum*" (Durkheim 1972: 70). The substratum was the desiderata of everyday life, practices, beliefs, laws—all manifestations of collective life,

> are expressly obligatory, and this obligation is the proof that these modes of action and thought are not the creation of the individual but come from

a moral power above him, that which the mystic calls 'god' or which can be more scientifically conceived.

(*Durkheim 1972: 73*)

Durkheim continued, "The function of a social fact can only be social.... We can thus [say]: *The function of a social fact must always be sought in its relation to some social end*" (74). Social life itself is a morally laden operation managed by reference to religious obligations which mirror the ideal (thus sacred) image of the social. As Durkheim wrote, "Thus it is seen that whatever has been done in the name of religion cannot have been done in vain: for it is necessarily the society that did it, and it is humanity that has reaped the fruits" (Durkheim 1915 [1965]: 467).

Social Solidarity and Religion

Tylor proposed a development of human societies grounded in Darwinian logic—the most primitive societies are outcompeted by more rational and complete explanations of the world provided by more advanced religions—with all religions eventually eclipsed by science. Durkheim, in contrast, found such a simplistic application of evolutionary theory to social structures problematic. Rather than a linear progression, Durkheim envisioned different categories of societies rather than a hierarchical placement of "civilized" societies above primitive ones (Durkheim 1972: 129). Instead, Durkheim proposed the world was comprised of two types of societies: those which operated in a state of mechanical solidarity, and those which exhibited an organic mode of solidarity.

Mechanical solidarity was characterized by direct, unmediated relationships, and exogamous modes of kinship—a segmented society in which each part was relatively similar, and shared a group conscience characterized by a mostly homogenous set of social norms and practices. Durkheim characterized individuals as "social molecules," which he said might "act together only in so far as they have no action of their own, as with the molecules of inorganic bodies. That is why we propose to call this form of solidarity 'mechanical'" (Durkheim 1972: 139). In such societies, *everything* is religion, and religion is a *moral* code written in the relationships evident in the habitat. These were, he said, the elementary forms of religious life (Durkheim 1965 [1915]). And in those cases, religion was deployed in ethically charged categorizations to channel human behavior. Religious practice was a system of etiquette for relating to other human and nonhuman animals. This was what he referred to as "politico-familial" form of social organization.

As societies grew larger, as personal anonymity increased, as laws took the place of socially mediated taboos, these segmented groupings blurred:

"to the extent that society becomes more concentrated, and these elementary groups lose their autonomy and become dissolved in the great mass, responsibility becomes an individual matter" (Durkheim 1972: 131). Here Durkheim described the transition from mechanical to organic solidarity. Rather than a grouping of similar and homogenous segments, organic solidarity should be imagined as the organs of a body, each performing its own special role, and which are in turn comprised of different social elements (Durkheim 1972: 143).

What drove the transformation from mechanical to organic solidarity was the specialization of labor, characterized by heterogenous new social elements and arrangements. Durkheim ultimately had a positive view of the potential of organically arranged societies, to the extent that the division of labor was spontaneous and liberatory for both individuals and society. More closely connected both temporally and spatially, an organic system also has a notable fragility. Whereas mechanical societies were comprised of parts which were relatively homogenous, organic solidarity's new social elements and structural arrangements created the conditions for *anomie*—the moral degeneration of society due to a perceived alienation of individuals from the group. Moreover, the emergence of widespread anomie was predictable to some extent, and might be ameliorated through appropriate interventions of the social unit. A properly attuned social body would foster "spontaneous," unfettered economic organization which would diversify the society in positive ways (Durkheim 1972: 182, 184*f*). "In short," Durkheim told his readers, "labor is divided spontaneously only if society is constituted in such a way that social inequalities exactly express natural inequalities" (182). In other words, the diversification of society into social several social "organs" challenged assumptions held in mechanical groups that "presuppose[d] the existence of this order, because, serving as a basis for the moral and religious system, it appears to them as sacred" (182).

The overarching moral code that served as the *conscience collective* became frayed, fragmented, and dissipated. In a society characterized by increasingly specialized social roles, the loss of a common vision or ethical standard was an obstacle. "To the extent that society becomes more concentrated," Durkheim reminded us, "and these elementary groups lose their autonomy and become dissolved in the great mass, responsibility becomes an individual matter" (131). No longer able to depend on the inherent social capital of the system, each person made moral choices independently, without the collective in mind. There was less accountability when one did not see or even know about the anonymous cascading effects of economic actions. The individual was left at the edge of the moral abyss with no dependable higher power upon which to call. Since a society in organic solidarity could not count on the conscience collective, a cult of the individual arose. Durkheim elaborated by saying "As all the other beliefs and all the other practices take on a less and less religious character, the individual becomes the object of a sort of religion" (146). He related these developments to the emergence of industry and larger and more specialized corporate organizations. The laws of society, Durkheim believed, governed the relationships between the specialized groups comprising organic society:

> since a body of rules is the specific form which is assumed by spontaneously established relations between social functions in the course of time, we can say, *a priori,* that the state of anomie is impossible wherever interdependent organs are sufficiently in contact and sufficiently extensive.
> *(Durkheim 1972: 184)*

In Durkheim's recipe, there was a central problem which lurked within the structures of organic solidarity. To draw on Durkheim's language above, a review of relevant research might lead us to conclude that keeping the organs *sufficiently* in contact and *appropriately* extensive is a tricky business. Just as subsidies distort markets in ways that reflect the values of certain social groups or corporations, the organs and the range of their influence can be manipulated. At this writing, corporations are engaged in manufacturing norms and values and spreading them in new and sometimes surprising ways.

If the reader has some exposure to the work of the political economist Karl Marx, they may perceive in Durkheim's logic some elements in common with Marx. For Marx, the mode of production—what he referred to as the structure—was foundational to and generative of the superstructure, which manifested as religion. Durkheim referred to this phenomenon as the division of labor, and both would agree that this had implications for religious complexes. They differ, however, on social alienation. Durkheim noted, referring implicitly to Marx and his ilk, that they "degrade the individual by reducing him to a mere machine" (Durkheim 1972: 178). The alienation Marx depicted resulted from a laborer's demotion from a craftsperson to a cog in a machine, with a concomitant loss of esteem and even physical proximity to others. Social accountability waned with the waxing of industry. For Marx, this was the obvious and logical outcome of industry. For Durkheim, such a situation was due to aberrant circumstances, what he called a corruption. In a society characterized by harmonious interactions between different social elements "the division of labor presumes that the worker, far from being hemmed in by his task, does not lose sight of his collaborators, that he acts upon them and reacts to them" (Durkheim 1972: 179). While both were methodological holists (they took the system as the unit of analysis), they had differing diagnoses regarding what caused the ills of industrialism, and divergent prescriptions. Both scholars pinned dramatic social changes on the effects of industrialization and specifically capitalist modes of production.

Each of their broad critiques pointed toward the ill effects of industrialization, a premise familiar to scholars of religion and nature.

There were further differentiations between the two thinkers. Durkheim was clearly a social constructionist—he ultimately believed that collective *ideas* were formative for religion and society (which are the two sides of the same coin). Marx, on the other hand, worked within a historical-materialist framework. That is, in Marx's analysis social change is caused by shifts in the mode of production (i.e., the specialization of labor). According to Durkheim, the *conscience collective* was the way to understand social change over time. Although he was an idealist, Durkheim rescued social constructionism from complete relativism by tethering it to a sort of "natural law" explanation for the cross-cultural and fundamental characteristics of society. Essentially, spirituality served the social function of "reaching individual consciousness of giving them a direction and of disciplining them," with the aim of paying out into behavior that was ultimately beneficial for the broader society (Durkheim 1965 [1915]: 467). More will be said about how Durkheim viewed capitalist industrialism, but for now, it is helpful to recount another feature of his theoretical analyses that was pertinent for religion and nature scholars.

Durkheim showed minimal interest in the landscapes where his target societies dwelled. When he did talk about land it was from a human-centered, utilitarian standpoint (what can that tree do for me?). Places were the source of the community's self-perception, and they were the theater in which antisocial behaviors were constrained by totems and social taboos. The construction of a sacred geography is a mere function of the society itself: "Far from being the case that society models itself upon the land, it is the land which bears the imprint of society. Thus, it is not the land which explains man, it is man which explains the land" (Durkheim 1972: 88). For Durkheim, the land was of secondary importance. Mircea Eliade, who based some of his fundamental presuppositions on Durkheim's work, would take a much different perspective on sacred landscapes (see Chapter 3).

When it comes to the study of the religion and environment nexus, Durkheim was important because the so-called primitive religions he described were environmental religions. For Durkheim, the important social facts elaborated above are "external to individual minds considered as such, in the same way as the distinctive characteristics of life are external to the mineral substances composing the living beings" (Durkheim 1972: 70). Durkheim's goal was to differentiate social from mere psychological facts. Some basic social facts are also, in our present age, "ecological facts," since the biophysical limitations of the planetary system are recognized by civil society movements, governments, and corporate entities across the globe (at least in letter). These socio-ecological facts constrained human behavior and development. Nonhuman materialities shape social relations; greenhouse gasses are planetarily agential (LeVasseur 2021: 24). They are basic facts of life today that were not realities at the time of Durkheim's writing. The academy itself, for example, is wedded to petroculture:

> As academics we need to understand that our entire professional project is scaffolded upon and enabled by use of dense forms of old sunlight. This literally means that academia as we know and practice it is by bio-ecological and biogeochemical definition, unsustainable.
>
> *(LeVasseur 2021: 63)*

The complicity of academic institutions in the distortion of the relations between the corporate and the social elements of society is evident—it is a contemporary rupture between the *social organs* posited by Durkheim. This was the lurking inconsistency in Durkheim's project: when social organs and their moral codes created conflict and anomie, this was considered an aberrant, abnormal state. Spontaneous forms of social action were imagined as the salve for this eventuality. But can such a strategy work when corporations are treated like sovereign citizens, while many humans go without access to reliable sources of food and water? Some thinkers influenced by Durkheim focused on the potential for disruption and danger that inhered in major social transitions. Before we get too far down the trail, however, it would be remiss to gloss over another influential sociologist of the twentieth century, Max Weber.

The Sociology of Disenchantment and the Iron Cage

Max Weber (1864–1920) initiated a sociology of religion that emphasized methodological individualism (in contrast to Durkheim), but crucially also insisted that his own research was comparative. Like Durkheim and others before him, he highlighted specialized production as one driver of religious shifts. The "spirit" of modern capitalism—its core ethos—promoted the systematic maximization of efficiency. At the same time, capitalism worked to maximize the rationalization and bureaucratization of society through the ethos of Protestantism, which emphasized individualization (direct, personal access to God), and efficiency (as manifested in some early movements) (Weber [1930] 2005).

The first Protestant movements, such as the one spearheaded by John Hus in the late fourteenth century, or Martin Luther's later protestations, frowned upon ostentatious public displays of wealth. Hussites (as followers of Hus were known) and later Anabaptists from central Europe retained a community-minded ethos but rejected the rigid hierarchy and lavishness of the Catholic Church. Others, such as those who followed more closely on Luther's heels, endorsed a more radically individualist (as opposed to community-based) ethos. To divest priests of their power over citizens, some of those early Protestants emphasized the soteriological importance of grace over works (reversing the Catholic Church's position), and focused on a personal relationship with God (rather than one mediated by a priest).[1] The personal and affective relationship with the divine was the theological seed that, when exposed to the vast and largely rural populations of the expanding United States (Turner 1920), would flower into the highly infectious religious movements associated with the First and Second Great Awakenings in North America (Hatch 1991).

Weber's preeminent contribution to the study of environment and religion was the analysis of this latter religious form associated with individualism.[2] Weber noted how one particularly strong mode of Christianity—Calvinism—as well as those manifestations of Christianity which were particularly important in the development of the United States (e.g., Puritanism), provided the widespread psychological conditions for capital accumulation. Calvinism and Puritanism both emphasized the notion of an omnipotent, omniscient deity who knew, *prior* to an individual's birth whether that soul was destined for heaven. Although God might know who was destined for heaven, such theologies fueled a pervasive anxiety among many Christians worried about their own individual souls. This anomie, Weber believed, combined with a capitalist ethos which promoted rationalization, modernization, and individualization. The rapid spread of this cultural cocktail spread rapidly in Europe and North America. From a theological perspective, the only way to tell if one was "favored" (predestined for heaven) was to see the fruits of righteousness manifested in *this* world through personal success or well-being. Likewise, the aim of capitalism was to maximize personal success and well-being. This was what Weber referred to as an "elective affinity" between the Protestant ethos and the spirit of capitalism (Weber 2005).[3] Importantly, however, while wealth was considered evidence of divine favor, publicly playing with this wealth was frowned upon. As Weber put it:

> asceticism looked upon the pursuit of wealth as an end in itself as highly reprehensible; but the attainment of it as a fruit of labor in a calling was a sign of God's blessing. And even more important: the religious valuation of restless, continuous, systematic work in a worldly calling, as the highest means to asceticism, and at the same time the surest and most evident proof of rebirth and genuine faith, must have been the most powerful conceivable lever for the expansion of that attitude toward life...called the spirit of capitalism.
>
> *(Weber 2005 [1930]: 116)*

Investment and re-investment of the profits from business ventures—the process of capital accumulation—demonstrated God's grace. It also meant that capital and power could be amassed without ostentatious clerical hierarchies, and expansive palaces. This material world was a fallen world for those American Protestants. The emergence of modern capitalism was thus driven by, or at least buttressed by, theological developments. Weber believed that the result of the Protestant ethic was the rise of modernism and the loss of a sense of enchantment. Although Weber's work had explanatory power, it was clear that the North American and European continents were not homogenous. Those who wielded increasing economic power through the colonial period, though, did share some theological and practical commitments. There were some movements or groups which do not entirely map onto the "ideal" Protestant in Weber's model, and these nuances were important complementary notes which complemented Weber's hypothesis.

Consider for example the Moravians, who ultimately traced their theological genealogy back to John Hus. The Moravians funded their extensive missionary work across the globe by exploiting exchange networks in North America, the West Indies, and Europe. In the late 1750s, the Moravians created the Commercial Society, which combined private and Church capital, cementing the bond between their religious vocation and transatlantic commerce (Gillespie & Beachy 2007). This group provided a provocative illustration of Weber's thesis that religious ideologies can play the role of a switchman (Weber 2005), actualizing some historical outcomes, and foreclosing others. They also challenged, however, Weber's assumption that rationalization, bureaucratization, and individualization resulted in a spreading cultural homogeneity. The Moravians founded their important colonial-era settlements in Europe and the United States on distinctly communitarian and collectivist values, in some ways radically opposed to the individualistic ethos of other mainstream Christian groups. But they also participated as much, if not more fully, in the transatlantic economy than many of their contemporaries. They thus contributed to the spirit of capitalism without embracing all the features of Weber's Protestant ethic. The story, then, was more nuanced than Weber imagined, though despite different values and motivations, there were many constituencies (including but not limited to mainstream Protestant groups) who shaped the ways in which capitalism and capital accumulation proceeded in the modern period.

While Weber's work was helpful in illuminating the relationships between religions and social developments, there were also geographic and demographic pressures that were largely outside his purview. The European Renaissance, indirectly and perhaps ironically spurred by several attempts to capture an important sacred space (Jerusalem), generated an increasingly wide-reaching series of exchange routes and fueled European cultural elites' appetites for faraway goods. Political and religious leaders shepherded the search for luxury goods and cheap resources to feed the growing political economy of Europe.

By the Renaissance era, Europe and Asia had developed densely populated urban areas, and in Europe, burgeoning populations often lived in close proximity to domesticated animals. Dense population centers facilitated the rapid mutation and transmission of pathogens (many of which were first transmitted from non-human to human animals), which significantly impacted Europeans. Their most visible and widespread effects, however, were among the Indigenous cultures that were the victims of colonization. As soldiers, priests, and bureaucrats were unleashed across the oceans in large vessels drawn from budding European industrial areas, they mapped, named, and then transformed the landscapes they encountered (Axtell 1992: 72). The first zoos and botanical gardens appeared in Europe to showcase the creatures (human and non-human) gathered from far-away lands (Chidester 2003). Bureaucrats laid claim to lands by cataloging and re-naming the world, a process which reflected the "encyclopedia" mentality of Renaissance Europe. To illustrate, consulting the royal historiographer Elio Antonio de Nebrija, Queen Isabella of Spain was supposedly informed: "Your Majesty, language has always been the companion of Empire" (Sale 1991: 18).

The geographic expansion of European cultures during the age of colonization extracted not only metals and riches but also important food sources, trade goods, raw materials, and medicines (Sale 1991: 46), changing the world's biophysical (and thus religious) landscape in significant ways. The capitalist mode of production (including the global expansion of markets, and significant increases in manufacturing efficiency, consumption, and wealth disparity) was constrained by geographic and biological features of the environment. Those habitats in turn shaped the theology and practice of religious individuals and groups. For instance, the unique communitarian ethos of the Moravian Church weakened and eventually dissolved as their settlements transformed from utopian frontier communities into centers of commerce and industry. The group grew into a relatively mainstream denomination that today, in many ways, is virtually indistinguishable theologically from other mainstream Christian groups. It is clear, then, that the affinity between the individualistic Protestant ethos and a form of capitalism which maximized rational pursuit of self-interest and promoted bureaucratization sometimes eroded the counter-hegemonic energies of smaller groups, bending them toward the political and social center. The emergence of social movements that were primarily rural and non-hierarchical spread through populations like a contagious cultural virus and moved into the mainstream as capitalistic individualism became its own global religion (Johnston & Wall 2014; Loy1997).

Weber predicted that a truly global mode of capitalism would transform and calcify socioeconomic relations into a self-perpetuating rationalized, efficiency-maximizing state. Drawing on Richard Baxter, an expert on Puritan ethics, Weber noted that in Baxter's view, the care for material goods should only lie on the shoulders of the worshipper "like a light cloak, which can be thrown aside at any moment" (Weber 2005 [1930]: 123). But fate decreed that the cloak should become an iron cage. This iron cage was the cultural inertia of capital accumulation through investment in a system that maximizes individualization and bureaucratic development, coupled with an unwavering faith in capitalism. This is where we find pastors who own multiple private jets and vacation houses or, like an unlucky self-styled bishop in 2022 who was robbed of over one million dollars in jewelry on camera while his program was being filmed.[4]

Disruption, Danger, and Mitigation

Religious social formations were environmental religions to the extent that they also had an important geographic element—they were social constructions specific to the habitats in which they took shape (Durkheim 1965 [1915]). This was important to understanding the scholar of religion Mircea Eliade's arguments, for Eliade radically departed from Durkheim's understanding of geographies of difference (see Chapter 3 for more on Eliade). But some of Durkheim's other intellectual progeny added nuance and methodological clarity to the field. Social conflict was a crucial and common element in the thinking of Durkheim, Marx, and Weber. For Marx, a historical materialist, the ongoing conflict was between the bourgeoisie and the proletariat—the working class and the ruling class. For Durkheim social conflict was also inevitable, but on balance was positive and contributed to social cohesion. Weber analyzed conflict in a more situational framework, attending to specific historical circumstances rather than perpetuating the preoccupation with religion's origins.

Social scientists who came after Durkheim left behind some of his colonialist baggage and advanced the study of human societies by focusing on conflict and danger. Victor and Edith Turner were anthropologists who contributed substantially to teasing out the *structure* of conflict in societies. It is worth noting that most works about their anthropological contributions mention only, or primarily Victor. Although Victor Turner got much of the credit for their work, they made it clear that Edith was always present in their fieldwork, and behind all the written work. Credit should properly be afforded to both. Their fieldwork in Africa made a prediction about conflict in small-scale groups (those that Durkheim would have referred to as examples of mechanical solidarity). The Turners posited that ritual action was formative for the solidarity of a group, and recognized that the point of ritual was to navigate potentially dangerous or polluting times. Rites of passage for women and men, marriage, death, and any other event that required the constraints of ritual had a *liminal* phase. Imagine it this way: the social group had a "normal" structure which was maintained at most times of the year. During some of these especially fragile moments when ritual action was required, the social structure was temporarily overturned: girls became women, two people were joined in union, a child was born, or a grandfather died—these are all passages within and between lives. These liminal phases—the anti-structure which appeared as the normal structure was made explicit in ritual, and then systematically overturned—were the key. The moment of *communitas*, or spontaneous community, that follows the liminal phase is the engine of social solidarity. The momentary suspension of the normal community functioned to diffuse tensions and heal the whole stronger than before.

The anthropologist Mary Douglas (1921–2007) argued that pollution and other dangerous circumstances helped communities redefine and/or sanction the boundaries of what, in that context, was taboo. Douglas thought that Turner had the process right, and he had generated a dynamic anthropology. But, Douglas corrected, those moments of what Durkheim called *communitas* were primarily marked by danger and uncertainty. In essence, where Durkheim was ultimately the root of structural functionalism, the work of Turner and Douglas (along with others such as Franz Boas) introduced a move toward historical particularism that was absent in earlier works. In other words there was a move away from grand narratives of religious origins and toward more contextualized research. Douglas challenged the dominant functionalist perspective and argued that pollution was a moral (not a functional) category. Looking at religion only through functionalist lenses was misleading, Douglas believed. Religion was about categorizing reality and defining (and redefining) boundaries. Pollution and religion were also always tied up with natural symbols (Douglas 1966, 1973), whether specific species or landscapes.

One of Douglas's important contributions for the scholar of religion and nature was the classification of societies as possessing the characteristics of what she termed "Group" and "Grid." Group referred to how a particular society related to outsiders, to other groups. Grid referred to the internal relations within the group and between individuals. Imagine a graphical representation of Group/Grid where societies that have a strong Group/Grid dwell in the lower left corner, and those with weak Group/Grid occupy the top right of our "ideal space." In societies with a strong Group/Grid, in-group and out-group exchanges are highly regulated and have internal mechanisms for diffusing conflict (social taboos). On the other hand, societies with a weak Group/Grid exhibit a high level of individualization and have few fetters on internal relations or out-group exchanges. Because of their heterogeneity relative to strong Group/Grid societies, those with weak Group/Grid are highly conflictual, lacking strong social structures for effectively mediating internal and external relationships. Note here the structural similarity between what Durkheim called mechanical solidarity and Douglas's strong Group/Grid cultures. But Durkheim imagined these social units as relatively undifferentiated. In her work, Douglas recognized that those societies were highly structured, and by recognizing this structure that Durkheim glossed over, it was possible to pinpoint the ways in which communities mediated and mitigated pollution and danger, through interactions of natural limits and established social norms.

Using Douglas's model, imagine that the United States is an example of a society that exhibits weak Group/Grid. On the other hand, groups with strong Group and Grid come in a variety of forms, from traditional communal groups, to diffuse networks of belonging with highly regulated relationships, which resist what they perceive to be the imposition of individualism and the erosion of traditional relations. These contestations over the erection and patrol of boundaries follow ecological patterns. The mode of subsistence in specific ecosystems, the natural limits imposed on those landscapes, and social arrangements weave a rich tapestry which tells who is "in," and who is "out" of biocultural boundaries.[5] With such a framework it is possible to imagine the United States as a hegemonic power grounded in the capitalist mode of production, and individualistic interpersonal relationships and obligations—weak Group/Grid. It is also possible to perceive that there are social identities that posess strong Group/Grid, whose identities are endangered by individualization and capitalism. Understanding relationships between human modes of subsistence, resources, and capital is key in considering the nexus of religions, cultures, and perceptions of nature.

The study of religions and the environment, then, is not just about how this or that particular religious group's texts or practices can be 'greened,' but also about how politics is reflective of socio-ecological factors and related directly to religious perceptions of how humans relate to nature. One recent illustration of this in the United States might be contestations over land between rural landowners and the federal government. Cliven Bundy, a rancher in Oregon, illegally occupied federal land on several occasions, challenging what he perceived to be an out-of-touch set of bureaucratic controls which impeded on their individual liberties (B. Quammen 2020). The rural populations including Bundy and other ranchers shared an ideology grounded in personal liberties, small government, and in this case, buttressed by a Christian nationalist identity formation. The lenses they look through—their identities—characterize impositions of governmental authority as challenges to their own values. These modern-day cowboys could be imagined as having a strong Group/Grid, resisting the imposition of hegemonic values over landscapes which they perceive to be rightly theirs. Armed conflicts and wars are contestations over access to and distribution of resources, and they are some of the most ecologically damaging activities in which humans engage. Perhaps Durkheim's sociology provides some theoretical grounding for how we might imagine, and be receptive to others with different social patterning, especially when it presents an occasion for conflict.

QUESTIONS FOR DISCUSSION:

- How did Durkheim attempt to distance himself from those who applied biological evolutionary theory to social groups?
- What caused the emergence of anomie in Durkheim's theory?
- Do you perceive any relationship between Durkheim's concept of anomie and Weber's notion of the "iron cage"?

Notes

1 The notion that grace is more important than works was related to the idea of predestination. If God is omnipotent and knows before an individual's birth the fate of their soul, then one cannot simply earn their way to heaven by doing good deeds, because that would make God's love contingent on human behavior, which raised questions about omniscience. Only divine grace, so the logic went, could be responsible for saving a person's soul.

2 Weber believed that the result of the Protestant ethic was evidence of the loss of a sense of enchantment in the profane world, and of the rise of modernism.

3 The translation by Talcott Parsons, cited below, translates the term "*wahlverwandtschaften*" as "correlations" rather than "elective affinity," which appears in the translation from Baehr and Wells. Thanks to Bernie Zaleha, who pointed out this difference in translations.

4 See https://www.cnn.com/2022/07/25/us/brooklyn-pastor-robbery/index.html, accessed 27 July 2022.

5 LeVasseur's book (2021) was a provocative elaboration of how the boundaries of our academic studies are conditioned by material factors. The contemporary academy in North America and Europe, for instance, is entirely dependent on ancient pools of carbon.

3

MIRCEA ELIADE

"Religion Is Real, Place-based, and Irreducible" (Study of Religions)

The Irruption of the Sacred and the Foundation of the World

One of the most influential scholars of religions in the twentieth century was Mircea Eliade (1907–1986), whose comparative work fundamentally depended upon Durkheim's distinction between the sacred and the profane. Importantly for scholars of religion and nature, it was Eliade who famously declared that the sacred would irrupt into the profane world, making obvious the distinction between them. The site of irruption was an *axis mundi*, or world center, for those who recognized this sacred intrusion into the profane world (Eliade 1959). Where Durkheim failed by reducing religion to its social function, Eliade advertised a study of religions wherein religion scholars could provide special insight not available to those devotees of other human sciences (Eliade 1959). That is, Eliade said, religion was *sui generis*, in its own category, and might *only* be properly approached and interpreted by those who had specialized knowledge in this area.

At that time, phenomenology was a prominent method of analysis and interpretation. It is beyond the scope of this short work to detail the characteristics of such a wide-ranging methodology. Succinctly though, scholars who deployed this approach focused on the subjective experiences of persons and imagined them as reflections of the *numina*—the metaphysical elements which attended the phenomenon.[1] The dialectic between these two poles was the fuel that drove society, and religion was the tool for its authentication. Eliade was typically portrayed as a phenomenologist, but this view requires some nuance.

For Eliade the sublime or awe-inspiring *axis mundi* was not merely a psychic manifestation of feelings of fear and trembling in the individual, as it

DOI: 10.4324/9781003364801-4

was with Max Müller (1889) or Rudolph Otto (1925). If Durkheim reduced religious phenomena to social causes, the psychologist Sigmund Freud distilled it into an individual psychological phenomenon (Freud 1964; 1998). Another psychologist, Carl Jung, referred to awesome experiences with the natural world in his theory of archetypes—the biologically based, objectively real manifestations of the collective unconscious.[2] They all reduced religion to something else—psychology, social fabric, culture, etc. But Eliade was up to something different: Eliade advanced an ontology and geography of religion. He wanted to open the possibility that there might be a certain magic that dwelled in the comparative enterprise because it illustrated the inability of human language to capture all the elements of the sacred that irrupt into the world (Patton & Ray 2000).

Eliade's definition of religion was also beholden to the theologian Rudolph Otto and his description of the religious experience (Otto 1925; Eliade 1959: 8–10). He approvingly quoted Otto's experience of the sacred, variously terrifying, overwhelmingly powerful, majestic, mysterious, and boundless (Otto 1925). Much like Müller decades earlier, Otto and Eliade advanced the idea that an ineffable experience *with* or *in* nature was the wellspring of what we might refer to as "the sacred." In his famous book *The Sacred and the Profane* (1959), Eliade defined the sacred: "The first possible definition of the *sacred is* that it is *the opposite of the profane*" (Eliade 1959: 10, italics in original). But how do we know the difference? Fortunately for his readers, Eliade went a bit deeper than this circular definition and asserted that the sacred becomes obvious because of a qualitative difference with the space around it afforded by the manifestation of the sacred in a geography—a *hierophany*.[3] This was typically observed in natural objects, including trees, rivers, mountains, and other outstanding features of the landscape. All religions and rituals, then, were the products of place-specific hierophanies. A hierophany might also occur through a holy person (a la Jesus of Nazareth in the role of the God), but it retains its strong geographical components.[4] "In other words," Eliade told us, "for those who have religious experience all nature is capable of revealing itself as cosmic sacrality" (Eliade 1959: 12). But hierophanies are not arbitrary markers in physical spaces. Rather they break the homogeneity of space and create the "revelation of an absolute reality… [which] ontologically founds the world" (Eliade 1959: 21). The hierophany not only orients humans to space, but also to sacred time.[5]

Eliade highlighted the use of the notion of *Terra Mater*, or Mother Earth (Eliade 1959: 138–151) among many cultures in the world, and noted that women were naturally attuned to their habitats (Eliade 1959: 144–147).[6] Mother Earth was personified as a single living organism. But Eliade extended that sense of planetary belonging to conclude that "one thing seems clear beyond doubt: that the cosmos is a living organism, which renews itself periodically" (Eliade 1959: 148). For Eliade, as for Tylor, Durkheim, and others, nature was a storehouse of symbols. But for Eliade, it was also some *place,* where the sacred physically manifested in a hierophany. It required ceremonies that recapitulated the cosmogony—the creation of the universe. A hierophany imbued a sacredness to place, and that sacredness was ontologically *real.* The construction of the world around that sacred center, the world axis, was a geographical and environmentally religious vocation. That raises an important question, then, that Eliade did not attend to, although it certainly would have been visible to him at the time had he followed his claims to their logical conclusion. The lingering question is *who* authenticates a particular hierophany as world-founding? What did that person or group have to gain from constructing the primary point of spatial and temporal reference in the world at *that* place and time? Groups disagree on what is a sin, and what is sacred. So which group's sacred geography trumps others? For instance, the Black Hills are sacred to the Lakota. A portion of the Black Hills has also become Mt. Rushmore, a symbol for the American nationalist civil religion. Does one group's sacredness supersede another's, or might they co-exist? Answers to these questions are unclear. Moreover, by invoking Jung's notion of archetypes Eliade had "biologized" the sacred: it follows that *"every construction or fabrication has the cosmogony as paradigmatic model.* The creation of the world becomes the archetype of every creative human gesture, whatever its plane of reference may be" (Eliade 1959: 45). Religion is the result of an "unquenchable ontological thirst...a [thirst] for *being*" (Eliade 1959: 64).

The "discovery" of agriculture, according to Eliade, fundamentally changed human relationships with their habitats. That formula may sound familiar to those who study the intersection of religions and environments. As human economies shifted so did their economies of the sacred (Eliade 1959: 126), resulting in the desacralization or secularization (Eliade used both terms in his work) of nature. At times, the sacred was "provoked," often with the help of other animal persons: "a sort of evocation is performed with the help of animals; it is they who show what place is fit to receive the sanctuary or the village" (Eliade 1959: 27). The same is true with other carbon-based organisms. "It could be said," Eliade insisted, "that all trees and plants that are regarded as sacred...owe their privileged situation to the fact that they incarnate the archetype, the paradigmatic image of vegetation" (Eliade 1959: 150, again with the nod to Carl Jung).

Yet, toward the end of *The Sacred and the Profane*, he dismissed what are often today referred to as "nature religions," those that contain elements of hybridity or which are just the irresistible urges of *Homo religiosus* that persist even among the "irreligious" (205). Eliade opined that these disappointing modes of religious production may still fall under the rubric "religions," even if they "almost always present the aberrant aspects of 'pseudomorphs'" (206). Nature religions look like authentic religions (pseudo-) and are shaped like authentic religions (morph), and may be imagined as encompassed by the category religion, provided we cast the net wide enough. But those pseudo-morphic nature religions were sycophants which, although not properly religious, still demonstrated an ontological thirst. They still needed a formative *axis mundi* that looked like, or operated as an anchor for sacred geographies. This is a good time to return to the above question regarding *whose* hierophanies provide the hook around which a society's myths hang. It is worth noting here that some of the most environmentally beneficent spiritual formations at this writing are incubating in dark green streams of religion which consider either the planet or preternatural spirits or forces related to nature to be a source of moral authority (Proctor 2005; Taylor 2010). And as Eliade himself noted, the social movements that dwell on the cultural periphery, the cultic milieu, are often hybrids cobbled together from multiple sources of authority and contrasting moral sentiments (Campbell 1972; Douglas 1975; Levi-Strauss 1979; Taylor 2002).

For some time, those who advocated leveraging the world religions to combat environmental decline mirrored a social scientific tendency to use forced-choice survey questions as categories for religious belongingness and practice. Just getting survey participants to choose one (and only one) religious identity, distilling the brimming diversity of religious cultural production down to the largest religions, or those global in scope, was always difficult. Even *inside* religious traditions, people disagree. Although Pope Francis's encyclical *Laudato Si* made a big splash in Catholic circles, it fell short of some of its biggest expectations. Seven of the Republican nominees for President of the United States in 2016, all Catholics, uniformly addressed the document in a debate by rhetorically relegating religion to the private sphere. The Pope, they insisted, did not make economic policy. In the United States, the religious sphere remained private, while public business was conducted according to capitalist principles. But for several decades in the twentieth century, the University of Chicago, where Eliade held a professorship, trained significant numbers of scholars of religions steeped in Eliade's "history of religions" approach. Toward the end of the twentieth and into the twenty-first century, criticisms of his approach appeared, highlighting some problematic elements of his theory. The scholar of religions J.Z. Smith was one important interpreter and critic of Eliade.

A Political History of the History of Religions

Smith called attention to deficits in Eliade's research program. Specifically, he interrogated the sacred geographies at the center (wordplay intended) of Eliade's theory. Rather than imagining the sacred impinging on the world as Eliade imagined, Smith described the construction rather than the discovery of the sacred. Smith's work *To Take Place* (1987) deconstructed Eliade's observations, noting specific instances where the supposed center—or *axis mundi*—was not a native category. In other words, it was not a term or concept that was drawn from the groups under study, but rather a product of the scholar's interpretive framework (Smith 1987: 15–16). We should not proceed as though what religion *is*, is a settled question. Rather, we ought to view each deployment of the term as related to the questions that a specific scholar is asking. As one example, Smith recounts the "sacred mountain" trope that Eliade perceived as a common thread running through many ancient religious complexes during the Second Urbanization (first millennium BCE) in the fertile crescent (Smith 1987: 16). It was easy, Smith reminded his readers, to project non-empirical commonalities on people and places long gone. The lesson here is that students of religions should be cultural detectives who investigate the deployment of categories related to the sacred. The observations of scholars tell us as much about the scholars themselves than they do about the objects of their study, because the questions scholars seek to answer are theirs, and not those of their interlocutors. Oftentimes narratives about cultural origins, or descriptions of others concept of the sacred, were products of the Western gaze, not objective or authentic renderings of that society's values.

Memories are used to construct and amplify values from the present that are then projected into authoritative figures, places, or things from the past, to justify current behavior. Recollections, Smith reminded his readers, are deceptive: we imagine memories as confined to the past, but they live at least as much in the present as the sense of the sacred is constructed through the lens of the past. "What if space," Smith asks, was "not the recipient but rather the creation of the human project?" (Smith 1987: 26). Spaces and places are not, contrary to what Eliade imagined, passively receptive vessels for the expression of the sacred. They are, rather, consecrations of particular (and sometimes peculiar) exercises of social, political, and social power. So, as Smith put it, "both temple and city can be seen as symmetrical or asymmetrical, depending on the perspective of the viewer. From the outside, the sacerdotal map is symmetrical, while from the inside, the priests do not enjoy equanimity" (Smith 1987: 69).

Theoretically speaking, why would grand theories of religion favor a *sui generis* experience of the sacred which is primarily cognitive and/or emotional? Why suppose walking a mile in another's shoes is one thing but understanding what is really in their mind is impossible? Durkheim, and Eliade after him had in common a presupposition prior to investigation regarding the nature of the human mind: that dualistic categorization is *natural* to human thought. The sacred and the profane, the pure and the impure are foundational categories for early enthusiasts who believed they had a bead on what religion really *was*. But Smith reminded us that "there is nothing inherently or essentially clean or unclean, sacred or profane. There are situational or relational categories, mobile boundaries which shift according to the map being employed" (Smith 1993: 178). When scholars imposed their constructed perceptions onto the lenses that aboriginal (or primitive, savage, primordial) individuals and groups used, they became tools that had social, political, and economic effects. Smith called attention to this problem: "What troubles me is that these two portraits of the primitive—the nineteenth century negative evaluation and the twentieth century positive (even nostalgic) appreciation—are but the two sides of the same coin" (Smith 1993: 179).

Inspired by Smith's reflective analysis the scholar of religions Russell McCutcheon and others treated the History of Religions (what Eliade called his research program) as data, rather than an explanation of for religious experience. McCutcheon's first book, *Manufacturing Religion: The Discourse of Sui Generis Religion and the Politics of Nostalgia* (1997) criticized the development of the method associated with Eliade and the History of Religions research program. That approach cloaked the category religion in an ostensibly apolitical façade, behind which some fundamental assumptions were smuggled into the History of Religions. Namely that religious *experience* was primary, and explanation or description was secondary; and that because of the inaccessibility of other peoples' minds, religion was *sui generis*—it was its own category (Martin & McCutcheon 2012). Other critics who advanced the field in the late twentieth and twenty-first centuries demonstrated that classification and categorization were not innocent and that they revealed much if scholars could uncover the often-obscured hierarchical assumptions upon which they depended. The research of Tim Fitzgerald and others illustrated the term *religion* does not fit the categories employed by cultural, ethnic, or ethical "others," and recommended abandoning the category "religion" entirely (Fitzgerald 2003a, 2003b;, 2015). The anthropologist Daniel Dubuisson recommended using the phrase "cosmographic formations," a term with less baggage and, he thought, was more precise (Dubuisson 2003). Religion was no longer something set apart, of its own category, it was always already *in* culture, *in* politics, *in* economics (Arnal & McCutcheon 2013).

Scholarly production related to religions and habitats has been variously rendered as "religion and ecology," "religion and nature," "religion and the environment," "spiritual ecology," and other appellations. Are any of those terms more useful than the others? Those who defend the use of the term *ecology* did so because it signified holistic, interdisciplinary, and multi-variate solutions to the ecological crisis. Most ecologists, however, likely have a more specific definition dealing with the behavior of communities of individuals and the interdependence of these communities. Scholars of religion Whitney Bauman, Kevin O'Brien, and Richard Bohannon addressed these questions in their book *Grounding Religion* (Bauman et al. 2011: 20).[7] *Nature* of course had its own complex and variable history. Was it a construct derived from a dualism between the cultural and the natural? Philosophically, did the use of the term nature allow us to dig even deeper by inviting us to think about the historical understandings of *nature* and how they are utilized today? Or is the term too broad to be of use? The same critical questions that apply to religion are relevant when it comes to nature. As soon as the distinction between the sacred and the profane came on the scene, there were questions related to purity and impurity. Is impurity imparted by "unnatural" uses of objects? Once despoiled by our species' exploitation does nature lose some of its authentically natural state? Do humans live *in* or *on* the world? The earliest scholars of religion puzzled over such questions.

The term environment also presupposes a certain dualism between humans and their habitats (rather than, say, viewing humans as part of our habitats). Methodologically, much of the work on religions and nature preserved presuppositions that religion operates in the head and heart (that it is a matter of faith), rather than in the hands (in the performative and socially relevant practices) of humans. In any case the term we choose may imply a focus on only global religions, or it may describe a sort of multiple religious belonging. Those presuppositions are embedded in the colonialist project that many of the early investigators of the category religion were. Namely, building and policing boundaries between those who evidence supposedly more evolved religions, and those who continue to operate within more homogenous and supposedly simpler social groups (recall Durkheim's mechanical solidarity). Statements related to the care and protection of the environment exist in nearly every religious group and denomination in the world. They are also represented in the Earth Charter, a remarkable vision which grew out of a series of consultations that aimed at being inclusive and democratic (Taylor 2010).[8] The consultations also presupposed that movement toward a global ethic for environmental care was a good and widely desirable pursuit. It is important to note that the same moral categories also lie behind neoliberal models of development which advance global-scale solutions to widespread problems. It is increasingly clear, however, that open and liberal societies based on maximizing individual well-being are often considered a poor substitute for locally held core values and religious belonging (Johnston 2014).

Digging for Data across Topographies of Power

Many Indigenous and traditional peoples publicly projected their own epistemologies to vie for political recognition and land sovereignty. Animals might advise, landscapes might heal, and water could run clear if somebody listened to what *they* need. The idea that the land has agency parallels some of Eliade's ideas about sacred space. Does that make Eliade's theories relevant for ongoing legal battles over land? At this writing, declarations and laws that grant that animals or ecosystems legal agency and rights have been introduced, and in some cases defended successfully in the courts (Chalee 2021). Claims about the sacrality of locales and species increasingly depend on a resistance motif that ontologizes sacredness. Is this a modern legal manifestation of Eliade's *sui generis* model of religion (Nobokov 2005)? Contemporary legal battles in the United States focused specifically on the sacredness of cultural sites in reference the UN Declaration on the Rights of Indigenous Peoples, and/or the UN Commission on the Human Right to a Healthy Environment. They have been invoked in several high-profile cases in Oak Flat, AZ, Bear Butte, SD, and in the Northwest Amazon (to name but a few). In each case, portions of landscapes have been declared sacred, and resources amassed to assist defenders of those places.

In Oak Flat there was a legal battle between Resolution Copper and the White Mountain Apache. The desecration of that land was noted and broadcast by Native tribes and their supporters. In this case (and in many others) they created alliances with local resource management agencies and independent ecologists who warned of the hydrological disruption that would result from the mine's development. Bear Butte in South Dakota was a site for ceremony and ritual for hundreds of years, and is sacred to the Lakota and the Northern Cheyenne. The butte was part of the land seized by the federal government in violation of a treaty signed with the tribes. For several decades it was in the hands of a private landowner, who advocated for protected status for the area. It is now contained within a state park. This arguably interferes with the Indigenous peoples' free exercise of their ritual lives since tourists and recreators can use the park at will. But the bigger insult for the tribes was the construction of an expansive full-service biker bar and concert venue on the edge of the park, very close to the butte. That area has the largest gathering of motorcycle enthusiasts in the world at the yearly biker rally in Sturgis, SD, only a handful of miles away. In both cases, these lands have been declared sacred to halt development. They do not declare the sacrality of these lands because they perceive in them an *axis mundi*—a world pole that orients their culture. Indeed, they have argued in the courts that they live in a sacred landscape which has many holy places within it, not just one. These sacred places are places of power, and whether or not they are energized by invisible forces (a non-empirical claim), there is empirical data about how people negotiate and contest the desecration of holy sites.

The anthropologist Robin Wright noted that the topographies of some tribes in the Northwest Amazon recognize a "world navel" within a sacred section of the river—what they consider to be their *axis mundi*—if one wanted to employ Eliade's term.[9] Of course, the tribes do not use Eliade's terminology. Wright criticized theories like Eliade's because they masqueraded as ahistorical and apolitical. In fact, the sacred geographies of many in the Northwest Amazon were not static, they changed significantly over time. Missionaries descended on those Indigenous worlds in the twentieth century. In some cases, tribes resisted the destruction of their lifeways. Charismatic strains of Protestant Christianity dwelled the jungle along with many other traditional spirits and powerful creatures. For the tribes, shamans maintained right relations with other persons (human and non-human) by tending to horizontal relationships. Christianity added a vertical dimension to the religious practices of the Baniwa and Aiary peoples (Wright 1988, 2009). Their God was now in the sky, but everyday life was navigated with the help of spirits and shamans. Some of Wright's work involved the characterization of Native Christians (Vilaça & Wright 2009), precisely the sort of multiple religious belonging that remains under scrutinized by scholars. So-called sustainable development projects which promised to bring the Baniwa and Aiary into the global economy resulted in preferential treatment of those who identified primarily as Christians (particularly Protestant-flavored ones) over traditionalist families. Changing the economy also disrupted traditional gender roles, and the transformation in the mode of production resulted in a transformation of the gods (Wright 1988). These changes precipitated a reactionary rise in assault sorcery in local communities among healers and shamans (Whitehead & Wright 2004).

While their cosmology and topography once reflected a sacred center and the orientation of the socio-political world of the Baniwa, they changed over time. It is not necessary to defend a claim about some *hierophonic* jump-scare (a la Eliade, Müller, and Otto) to analyze what is going on. Leaving aside the question of whether those places are actual or *real* irruptions of the sacred, we can still see an ontology of sacrality being wielded for legal, political, and social ends. That is, for cultural detectives interested in evidence, whether or not the nonempirical claim to authentic sacrality is true (we do not have to make a ruling on that!), *it is still true* that claims of the sacredness of specific places are deployed for material and empirically accessible ends. It is not only possible but important for students of religions and environments to take as data the ontologies and normativities of cultural, ethnic, and ethical others (von Stuckrad 2007; Johnston 2014b, 2015). Normativities are values claims, while ontologies are the perception of how the world *really is* for those individuals or groups.

Unfortunately, all-encompassing moral visions and geographically static identity formations are not attractive to those who resist what they perceive to be the homogenization—the McDonaldization—of the world (Atran & Medin 2008; Medin & Bang 2014). Rejections of globalism can be found, for instance, in the writings of radical religious movements in many contexts, and not just environmental ones (Taylor 1995). The lesson is this: presuppositions which underlie universalizing claims—whether about religion, or nature—should be conundrums for study (Ivakhiv 2008; Taylor 2007; Wright 2009). Global-scale convergence of values propositions need not be a precondition for preserving biodiversity and adopting a mode of religiosity which values non-human nature. Eliade's theoretical approach is no longer widely utilized in religious studies as an analytical tool. Glimpses of his motifs can be found in the popular work of Joseph Campbell, or in some streams of depth psychology, which incorporated Eliade's concepts into Carl Jung's work on archetypes and self-actualization.[10] Eliade's book *Shamanism* (1951) was adopted by hallucinogenic enthusiasts as validation of the spiritual dimensions of life exposed by sacred plants. It also suggested that shamanism was a relatively widespread mode of religion, and that the experiential elements of religions were universal. As the critics above argued, however, any claims that appear to be ahistorical, apolitical, and universal are good opportunities for cultural detectives to engage their bullshit detectors.

Eliade's contributions to the study of religions are difficult to overstate. His work was carried forth by many students who populated departments for the study of religions for many years. Responses to and critiques of Eliade's methods and conclusions occupied the academy for decades more. Eliade's contribution to the study of religion and environment is the ontologization of the sacred, which may have some parallels with how rights of nature movements have unfolded in the courts.

QUESTIONS FOR DISCUSSION:

- What were the differences between Durkheim's and Eliade's understanding of the sacred and profane?
- For Eliade, what was the purpose of religious rituals?
- Is Eliade's definition of religion ethnocentric? Why or why not?

Notes

1 The notion is that there is a world of *phenomena* and a world of *numina*. The term *numina* has a long history going back at least to Immanuel Kant (1881). Hegel believed the dialectic—the tension from the contradictions between the two worlds—consisted of "Matter" and "Spirit." Ludwig Feuerbach (1837–1872) argued provocatively that God did not make humans in "His" image (the gendered noun is intended). Rather humans projected their own ego assumptions into supposedly divine beings, powers, or forces (Feuerbach 1957). Feuerbach took the "Spirit" out of Hegel's model and made the dialectic one that was entirely materialist—he had no need for the *numina*. Karl Marx (1818–1883) approvingly cited Feuerbach's materialism, and proposed that the materialist perspective prioritized the "structure," which for Marx was the mode of production. The "superstructure" was a projection of human desires and shortcomings. The alienation of capitalism resulted in religious production that was a salve for the

masses of faceless poor persons, and a justification for their continued exploitation at the hands of the bourgeoise, the ruling classes who controlled the mode of production.

2 The collective unconscious was a part of Jung's architecture of the mind. In theory, everyone has a connection to the collective unconscious. The collective unconscious is populated by biologically ingrained concepts—*archetypes*—which manifest in various ways in the world. One archetype, for example, is the hero motif, where there is an individual who faces an existential challenge which must be overcome.

3 *Hierophany* literally translates as bringing the sacred to light.

4 Eliade thought that the historical person of Jesus (if indeed he existed) was himself a *hierophany*. But it is an interesting question (a rhetorical one at this point) whether Eliade would think that subsequent apparitions of Jesus carried the same weight as Jesus himself. Would they also "found" the world? Would the Shroud of Turin, which supposedly touched Jesus and was subsequently emblazoned with Jesus' likeness, be a *hierophany*? If so, then this is another irruption of the sacred into otherwise homogenous space. But how could this be considered the foundation of a worldview? Even more provocatively, what about when Jesus appeared in the window of a bank building, or in a piece of toast? These all depended on a process of authentication enacted around them, precipitated and enforced by those who wield social, political, and economic capital. Perhaps from this perspective the axis mundi is a product of political, social, and economic engineering.

5 The term *ontology* refers to a description of the world *as it really is*, what is actually there. So, to say that an axis mundi "ontologically founds the world" is a way of saying that the hierophany is the geographic *and* moral center of the world.

6 Later chapters will explore these assumptions in more detail, specifically the chapters dedicated to Carolyn Merchant, Rosemary Radford Ruether, and Vandana Shiva.

7 The volume is now in its second edition (2022).

8 See https://earthcharter.org/read-the-earth-charter/, accessed 24 August 2023.

9 Wright, personal communication.

10 Archetypes are supposedly universal themes which developed in specific cultural contexts. Jung believed that altered states of consciousness (through psychoanalytic or religious ritual) provided individuals access to that trove of motifs. Thus, the cross-cultural appearance of structurally similar myths, legends, ceremonies, and relationships is drawn from the collective unconscious.

4

ROSEMARY RADFORD RUETHER

"God Looks Sexist When Viewed Through His/Stories" (Ecofeminist Theology)

There were various ways in which scholars challenged the status quo—they might be categorized as efforts at (a) recovery, (b) redescription, or (c) revolution. Those interested in *recovery* found new or under-scrutinized glimmers of green in their own traditions in the quest for sustainable environmental futures. *Redescription* includes efforts to bend an existing religion toward a greener future. There was no environmental movement when many religious texts or communities were created. But religious institutions are in the business of making ancient teachings relevant for the present. *Revolution* is a process where existing religions are changed to such an extent that they are barely recognizable. Some new religious movements fall into this category. Rosemary Radford Ruether (1936–2022) was an ideal exemplar of the second form of criticism—redescription. She challenged long-accepted standards within a specific religious group. Specifically, Ruether was a Catholic, and she did not pull any punches when highlighting the historical complicity of the Church in the construction of empire, the denigration of the body in a dualistic focus on the spiritual, and the quite obviously sexist hierarchy of the Church. But she did not attempt to leave the Church. Instead, she leveraged her insider status to push the Church toward more life-affirming and sustainable theologies.

In her remarkably powerful prose, Ruether offered a critique of Catholicism and marked the emergence of a new planetary ethos. The new planetary awareness focused on eliminating patriarchy from the Church, and the re-inhabitating of our earthy bodies for a theology that affirmed the goodness of the created order. She was adamant that this required transforming human relationships to each other, and to the Earth. Ruther was crucially important to the intersection of religions and environments for three key reasons. First,

DOI: 10.4324/9781003364801-5

her work illuminated the deeply rooted oppressions in social structures all over the world. So, the scope of her work was global. Second, she focused on the oppression of women both in historical theology and in contemporary practice, and connected these with environmental precarity and concern. Finally, Ruether exposed the erasure of matrilineal lifeways and the ascendence of patriarchy in the spread of empire, especially the ways in which various forms of oppression were justified by poor textual interpretations.

Liberation Theology, Or, Theology from Below

Ruether's earliest work, *The Radical Kingdom* (1970), was a fascinating exploration of the idea of *messianic hope*. She traced its transformation from Christian roots to secularized uses in, for instance, the French Revolution and 19th century Marxism.[1] The actualization of the messianic hope was sometimes imagined merely as an inward journey of self-exploration and discipline which concluded in a *spiritual* immortality *above* and *beyond* this finite, geophysical world. Ruether drew a correlation between the emergence of messianic language in theology and the use of a secularized form of messianism apparent in the post-Enlightenment separation of the Church and the political bureaucracy. Ruether also noted that the emergence of capitalism and the Protestant Revolution reinforced the separation of the material from the spiritual, and assumed an economic model where humans were bounded, discreet, cost-benefit calculators.[2]

Ruether was far from the only critic of the positive feedback loop between Christianity and capitalism (cf. Weber 2005 [1930]; also see Chapter 2 in this volume).[3] In many cases, the Catholic Church's position on economics was antagonistic toward both capitalism and socialism. Some papal encyclicals attempted to steer a path between the two.[4] One important twenty-first-century encyclical was *Laudato Si* (2015), which detailed the evils imposed on the poorest and most marginalized populations because of the degradation and pollution of nature. The Pope focused on the role of capitalism in controlling access to resources, and their liquidation for profit (*cf.* Pope Pius XI 1937; Paul, VI 1968; John Paul II 1981).

If we cast our gaze back to 1968, in Medellin, Columbia, the second Latin American Bishops' Conference released a document called *A Theology of Liberation* (Gutierrez 2023 [1973]) in which the marginalized and poor were the instrument through which God's redemption would be made manifest on Earth. But the Bishops also pointed to what was becoming increasingly obvious: the so-called Third World suffered from a systematic exploitation of resources justified by an other-worldly theology (Ruether 1970: 10). Such exploitation typically began by dispossessing already marginalized peoples of their land. Access to resources became the provenance either of the state, or, later, of corporations. Through bureaucratic fiat or privatization, the commons transformed—from a shared resource managed by local rules and enforced by local communities—into a commodity. There was a growing call in the Americas for *liberation* from the alienating effects of dualistic theologies, and the commodification of nature. Ruether contributed to and celebrated such movements, which proliferated through Latin America in the 1960s and 1970s. Ruether conceived of redemption as a material as well as a spiritual goal. Redemption was not an interior state or a promise of fealty to God. Rather, it was the actualization of a just and peaceful kingdom here on Earth. Ruether argued that "the apocalyptic view of redemption is basically social and outer-directed. One does not look inward to the salvation of some personal essence; one looks outward at history and society, at injustice, oppression, and cruel and irrational destruction" (Ruether 1970: 9).

Throughout her life Ruether demonstrated an interest in the emergence of countercultures, including the "New Left" and other associated radical social movements.[5] For example, she commended the revolutionary spirit discernable in actions such as Sit-ins, or the Be-ins. The sit-ins were demonstrations, often led by college students, where people occupied lunch counters reserved for White people in protest of segregation. Important to the sit-ins and the broader civil rights movement was a strong anchor in black theology and religion. The prose of Martin Luther King, Jr., the civil rights leader, challenged ostensibly supportive Whites to act—complacency in the face of injustice was a tacit assent to it (King 1963). He (and others of course) used the biblical rhetoric of exodus, freedom, and justice to rail against segregation. The Be-in was an event held in 1966 in San Francisco's Golden Gate Park. As a response to what they perceived to be a smothering rigidity in their parents' generation, countercultural movements exploded, especially on the West Coast, and in pockets in the Northeastern US (Fields 1992; Johnston 2006; Johnston 2014). In this rejection of the values of their predecessors, many in those varied scenes sought out alternative sources of authority. Many young radicals looked toward Eastern traditions such as Buddhism and Hinduism, and to Native American traditions, philosophies, and lifeways (Campbell 1972; Fields 1992; Taylor 2002; Taylor & Johnston 2016). Ruether recognized in the New Left and the hippie movements some affinities with some radical Christian movements:

> ...both the drug element and the sexual element in the counter-culture, if pressed to their deeper conclusions, lead to a renewed grappling with the problems with which all religious spirituality has struggled, namely, the problem of inward conversion, rebirth, and spiritual development that can

> lead the human soul from alienation and exploitation to a life style more in communion with both cosmic reality and the fellow human.
>
> *(Ruether 1970: 31)*

The overt and powerful spirituality that came with the countercultures was also realized in artistic developments, as well as cultural experiments in sexuality and cohabitation (Johnston 2014). Perhaps part of Ruether's attraction to these constituencies was that they were anti-authoritarian and non-violent (Ruether's 1970: 241–262). In the end, however, Ruether dismissed the sort of spiritual seeking associated with the countercultures as inauthentic because it relied on the use of drugs and other reality-skewing substances to provide a deeper connection and sense of fellow feeling with the Earth. Nonetheless, the anti-authoritarianism, justice-oriented values, and communitarian experiments with cohabitation she recognized there remained important elements of environmental resistance and philosophy. Clear already in her early writing was a strong stance against various intersectional forms of oppression.

Ruether's unique contribution, however, was her development of a theology of liberation from patriarchal false narratives, the gods of imperialism, and the power-laden exploitation of women and of the Earth. As Ruether put it in her important book *Liberation Theology: Human Hope Confronts Christian History and American Power* (1972),

> The gift of liberation, although alien and transcendent to the situation of sin, is not alien to 'our natures,' but springs from the same 'ground' as [humans'] original foundation. So, it is not properly seen as 'supernatural,' but as a restoration of [hu]man[s] to [their] true self, and a reintegration of creation with its true destiny as 'God's Kingdom.'"
>
> *(Ruether 1972: 9)*

Ruether was dragging the Christian ideal of repentance and liberation through the dirt, so to speak—re-materializing the notion of liberation, rescuing it from some dualistic valorization of spirit over nature. In Ruether's prophetic words:

> it is easy for those concerned with social justice to see the concern for ecology and reconciliation with the body as mere selfishness. But, in fact, oppression of persons and oppression of environments go together as parts of the same mentality.
>
> *(Ruether 1972:18)*

Remarkably in this early work of liberation theology, Ruether presaged later critiques of environmental movements from the global South.

Specifically, some scholars perceived that concern for the environment and especially wilderness was a luxury of the "First World," and that the categorical distinctions between spirit and materiality so fundamental to Western ideation did not neatly translate to other cultural worldviews (Guha 1995, 1996, 2000; see the Chapter 8 for further elaboration). Ruether had earlier noted that there were voices who perceived that social justice was separate from, and more important than ecological issues (Ruether 1972). According to the typology offered in the Introduction, those social groups were light brown, or at best light green.

As Latin American liberation theologians confronted political elites there, Ruether's brand of liberation raised the possibility of other emancipatory theologies. If the salvific message of Jesus was to manifest physically across the Earth, then liberation theologians surmised that each different marginalized group may have a theology of liberation that placed their own struggles against oppression in their proper sociocultural contexts. The Brazilian theologian Jon Sobrino offered the people of South America (and especially Brazil) as an embodiment of Jesus' suffering (Sobrino 1994). The freedom promised in the Christian Bible, Sobrino argued, was a this-worldly realization of the principles of justice (Sobrino & Ellacuria 1990; Sobrino 1994a, 1994b, 2008). In fact, any oppressed people might be imagined as a crucified people. A student of the famous theologian Paul Tillich, C.S. Song, translated the theme of liberation theology into an Asian context (Song 1996). In 1986 the important theologian James Cone published the seminal book *A Theology of Black Liberation* (Cone 1970). Ruether, however, was breaking new ground for this work more than a decade prior to these important contributions. Her chapter "Is There a Black Theology? The Validity and Limits of a Racial Perspective," illustrated one of the ways in which a Catholic theologian addressed racialized versions of liberation (Ruether 1972).

God May Not Be Sexist, but the Church Is

In 1974 Ruether published *Religion and Sexism*, which unpacked the role of women in Jewish and Christian traditions from all over the world. Another of her most impactful works was *Sexism and God-talk* (1983). There, she reviewed the emergence of patriarchal religions in the Near East, and the resultant Jewish and Christian monotheisms. Those patriarchies obscured matrilineal history by re-writing it, as men consolidated their political and military power. Her work shook the theological world. She wrote with power and urgency about the impacts of deeply seated androcentric (male-centered) hierarchies. Ruether tracked the mistaken Christian narrative that humans' natural place was exercising their dominion over the world and its persons (only some of whom were human), just as God ruled over the created cosmos as a divine king. That narrative's most dangerous outcome was the assumed dualism that separated humans from the rest of creation. As Ruether put it:

> All of the basic dualities—the alienation of the mind from the body; the alienation of the subjective self from the objective world; the subjective retreat of the individual, alienated by the social community; the domination or rejection of nature by spirit—these all have roots in the apocalyptic-Platonic religious heritage of classical Christianity.[6]
>
> *(Ruether 1972: 115)*

Although Mary the mother of Jesus played a smaller role in the Protestant interpretation of the gospels, she remained very important in the Catholic Church. Ruether extended analyses of Mary's role in early Christianity by focusing specifically on the Mary of the Bible, and Mary *in* the Church (Ruether 1977). As Ruether explained, at its very root Christianity was tainted by a neo-Platonic brand of Christianity, leading to a world in which the male became the default gender, heterosexuality the norm, and where White people exercised relational domination, which grew into systemic domination.[7] There was a glimpse, in the guise of Mary, of the ancient goddess religions that preceded Christianity. However, the male-dominated hierarchy of the Church only reserved space for women in the domestic, but not in the spiritual or political realms. The result was a myth that the divinely ordained male figure ought to be the head of the household (the *oikos*),[8] and women should be servants and child bearers. *Sexism and God-Talk* (1983) extended her analyses and directly connected the Mediterranean monotheisms to more ancient goddess religions. It was a head-on critique of mainstream Christianity—the first systematic ecological-feminist theology of Nature. Ruether constructed the image of a Christianity that was not bashful about the immanence of the divine. In other words, the divine, whatever was, was present in the material world, not beyond it. It was like drawing back the curtain to expose a millennium of man-god fetishes. For Ruether, Nature and God were co-existent—they were one and the same. God was relieved of the burden of gender stereotypes; an ecological feminist theology of Nature rejected "the model of hierarchy that starts with non-material spirit (God) as the source of the chain of being and continues down to nonspiritual 'matter'....[instead] The God/ess who is primal Matrix, the ground of being-new being, is neither stifling immanence nor rootless transcendence. Spirit and matter are not dichotomized but are the inside and outside of the same thing" (Ruether 1983: 85).[9]

If *Sexism and God-Talk* offered the first book-length explication of an ecological-feminist theology, *Gaia and God* (1992) advanced a moral sensibility grounded in this materialist ontology, an ethos of world healing. She wondered in that book whether Western science acted as a "new creation story," and compared religious and contemporary stories about the abasement of the world for the good of one species (humans).[10] And historically, to be honest, really only a small subset of that species. An ethic of healing, an ethic of care, was an embodied, empathetic process of being open to cultural, ethnic, or ethical others. Her theology was dynamic, and not fixed. It was poly-centric and turned the ethic of dominion on its head. It treated the Earth as the body of God (Ruether 1992; McFague 1994).

World Religions and the Environment

By the late 1960s and early 1970s, religious denominations began to elaborate statements regarding care for the natural world. The statements—some from Church hierarchies and some from well-regarded figureheads—advanced a concern for international interdependence ("we're all in this together!"). In these statements, they recalled religiously grounded narratives of hospitality for the "other," or care for the marginalized and poor—basically ethical regard for non-kin persons. Some of these statements and documents were responses to a short short five-page article in the journal *Science*, a widely read and influential journal, which catalyzed much of the environmental awareness within the Abrahamic traditions (Judaism, Christianity, and Islam) (White 1967).[11] The historian of technology and science, Lynn White, Jr., argued that the ecological crisis was directly traceable to the positive feedbacks between increasingly invasive agricultural technology and the Judeo-Christian dominion theme in the Hebrew Bible (the Christian Old Testament) (*cf.* Peterson & LeVasseur 2017).[12] While he opined that Americanized forms of Zen Buddhism, in particular, may be particularly eco-groovy, they were not native to the Western populations that precipitated the ecological crisis. Instead, White said, perhaps those from Abrahamic traditions might find resonance with the example of St. Francis (1181–1226), a monk from Assisi, Italy who it was said preached sermons to the wild creatures of the land and air. White even unofficially nominated him as an ecological saint. Two decades later, Pope John Paul II made White's idea a reality, declaring Francis the patron saint of ecology. At this writing, the current Pope is Jorge Bergoglio, who selected the papal moniker Francis to endorse the social and ecological justice represented by the monk Francis of Assisi.

In 1986 His Royal Highness Prince Phillip, President Emeritus of the World Wildlife Fund (WWF), convened a meeting of religious leaders to address growing environmental concerns. The meeting in Assisi, Italy (birthplace of St. Francis) included representatives from what were then considered the five major world religions: Buddhism, Christianity, Hinduism, Islam, and Judaism. (It is important to note that the composition of the "world religions" clique changes over time (Masuzawa 2005)).[13] From that meeting grew the Alliance of Religions and Conservation (ARC), the first non-governmental organization dedicated specifically to the intersection of the world religions and environmental care (Palmer & Finlay 2003; Johnston 2013). The *religious environmentalist paradigm* grew from this enlivening cultural loam. Poul Pederson and Arne Kalland used this phrase to describe a movement that generally adhered to a specific set of principles. First was the notion that religious worldviews *caused* eco-friendly behaviors. Once a particular set of ethical commitments was accepted intellectually, it would spur consonant behaviors. Second, Western, and especially Abrahamic traditions were grounded in dualism and the notion of human supremacy. Third, "other" religions, Asian, Indigenous, and contemporary animistic worldviews were inherently environmentally beneficent, promoting holistic perspectives and an ethic of interdependence and care (Kalland 2005; Taylor & Johnston 2016).

One impressive manifestation of the religious environmentalist paradigm was the series published by the Harvard Center on the Study of Religions on World Religions and Ecology (Tucker & Grim 1994; Tucker and Williams 1997; Tucker and Berthrong 1998; Chapple 2000; Ruether and Hessel 2000; Girardot et al. 2001; Grim 2001; Chapple 2002; Tirosh-Samuelson 2002; Foltz et al. 2003; Grim and Tucker 2014).[14] Following the Harvard Series, which in this instance included nine world religions, architects of the project, religion scholars Mary Evelyn Tucker and John Grim, launched the Forum on Religion and Ecology (FORE) to promote and diversify resources related to religious environmentalism. Their efforts included advocacy for a global scale religion and a minimum common denominator ethic. Motivated by the environmental philosopher J. Baird Callicott's suggestion that the world's religions should be mined for their conceptual resources for environmentalism (see the Chapter 8), Tucker and Grim spearheaded the group and were active emissaries of green religions. What they referred to as a minimum common denominator ethic depended on scientific foundations, and the good faith efforts by religious people to translate this world story, this universe story, to their flocks (Swimme & Berry 1992; Berry 2000; Swimme & Tucker 2011). To put it another way, cultural traditions varied of course, but to the extent that exemplary groups set aside their metaphysical commitments and adopted a cosmology derived from Western science, they would invariably be able to discern some basic, common ethical principles. Those within the religious environmentalist paradigm were a sort of new, ecumenical religious movement—scientifically grounded, conceptually inclusivist, and ethically cohesive (Taylor 2010; Taylor & Johnston 2016). Overall, the FORE was optimistic about the human future and about the potential role of religion in creating a more just and sustainable world.

Globalizing Ecofeminism, Cultivating a Planetary Ethic

Although Ruether was an editor of one of the Harvard series volumes on Christianity and Ecology (with the theologian Dieter Hessel), her published corpus also took stock of the struggles of women in the developing world (Ruether 1996, 1998, 2002, 2005), and always her critical eyes turned toward capitalism. For Ruether, globalizatoin by itself generated unjust social, political, and economic hierarchies. But the form exercised over much of the globe, Ruther said, was *corporate globalization*, where corporations manipulated or otherwise forced public policy into arrangements that were beneficial for businesses, but not for everybody (Ruether 2005). And those corporations were evangelical about capitalism. It turned out, corporate power structures proudly said, that *anyone* could support corporate globalization even if they were personally damaged by it, because consumer demands drove the creation of new products made primarily by corporate interests.

To combat corporate globalization, Ruether wrote that "whether religious communities see this self-sustaining ecology of nature as created by God or as the self-expression of an immanent divine being is perhaps less important than that they actively cultivate respect for nature's own life and see themselves as having to harmonize their own behavior with it" (Ruether 2005: 79). She was talking about human communities as the primary unit of moral considerability. That humans are the focus of human-created spiritualities seems unsurprising. But note Ruether's twist—that those human communities where Scripture is of primary importance must submit their own will to suit the ecological web of which they are a part. While humans are, ethically speaking, the most important creatures, any cultural practices that did not fit with the surrounding habitat would eventually die off because they were not adaptive in their ecological circumstances.

Present throughout Ruether's written corpus was an explicit disdain for the way in which capitalism systematically reduced cultural and biological diversity. Ruther's view was always global, but she connected her perspective with issues of race and class, not just within the Catholic Church, but outside it as well. Her work *Women Healing Earth* (1996) was an exemplar of her global vision—punctuated by the subtitle: *Third World Women on Ecology, Feminism, and Religion.* It was by highlighting the developing world that other luminaries such as Nobel Prize winners Wangari Maathai and Vandana Shiva (see Chapters 7 and 9) were given platforms from which to speak. They illustrated that women held important traditional knowledge that was often ignored by Western managerial approaches to resources.

With her materialist view of social change (influenced by Karl Marx's view of political economy), she brought theology down to the people so that theology *mattered* in the sense that it was embodied and performed, whether through suffering, acceptance, or resistance. If the *ecological* part of religions and ecology was taken seriously then planetary biodiversity was the matrix of our existence, the branch of the tree upon which we sat, as our exploitation chopped away the trunk of the tree. Such a view of the interconnectedness of life was, Ruther noted, historically feminine, maternal. The "matrix of being, which is not less the foundation of human beings, is rapidly depleted. Within two centuries this pattern of thought and activity has brought humanity close to the brink of the destruction of the Earth and its environment" (Ruther 1975: 194). She incriminated patriarchal societies ruled by corporate globalization, whose mode of production externalized environmental costs, which eroded biological and social integrity: "It is not too extreme to see this denouement [of the ecological matrix] as inherent in the fundamental patriarchal revolution of consciousness that sought to deny that the spiritual component of humanity was a dimension of the maternal matrix of being" (Ruther 1975: 194). Patriarchal consciousness manifested in oppressive social systems where women were objects meant mainly to accessorize men. Indeed, the relationship which perceived one or the other of a pair as dominant and the other as dependent was precisely the outcome of such a dualistic worldview. Therefore, "the structures of patriarchal consciousness that destroy the harmony of nature are expressed symbolically and socially in the repression of women" (Ruther 1975: 196).

Resisting Empire

Ruether's book *Integrating Ecofeminism, Globalization, and World Religions* (2005) advanced a strong argument that corporate globalization and capitalism focused on a top-down mode of globalization (managed by corporations with the assistance of financial institutions) without considering other possible modes of globalization. Ruether defined corporate globalization as "simply the latest stage of Western colonialist imperialism" (Ruether 2005: 1). Importantly corporate globalization was not some recent development, a sudden extension of corporate power over existing global networks. For Ruether corporate globalization must be considered deeply entangled in the context of 500 years of the colonial enterprise. The extension of monarchy-endorsed exploration and dispossession spread concepts of dualistic, hierarchical social arrangements in new colonial contexts. Even as the United States shook free of the monarchical yoke the ideas animating the creation of the colonies preserved those dualisms and grafted them onto a new, nationalized enactment of manifest destiny. The globalizing Western European and later North American imperialist power manifested in the scientific and ostensibly objective collection and classification of plants and people (Chidester 2003; McCutcheon 2019; Martin 2023). The first botanical gardens displayed the flora from faraway lands, and zoos incarcerated the world's fauna, reducing them to showpieces. People, too, were imported from across the globe and paraded in front of European and American audiences. Physical characteristics of cultural, ethnic, and ethical others were imagined as visible manifestations of their arrested evolutionary development. As Africa and much of Asia were carved up by European and American imperialism, crucial ingredients to development were the celebration of capitalism and democracy. The economic system (capitalism) and the political structures which facilitated it (democratic ones), it was imagined, maximized individual freedoms.[15]

The most recent and most damaging phase of neocolonialism unfolded following World War II (WWII). At the close of the war, several important institutions emerged that, together, would exploit the resources of the global South under the guise of free markets and free individuals. U.S. President Harry Truman unfolded a four-point plan to avoid any chance of another global conflagration. One of the elements of his plan was to help "underdeveloped" nations build economies robust enough to stand on their own. Although guided by the best of intentions, it turned out that *development* became a euphemism for surreptitiously increasing colonial dependence on Western aid (Ruether 2005). Following WWII, the United Nations (UN), formed from the shell of the impotent League of Nations (which had formed after World War I), emerged as the social-democratic arm of development. The UN linked the well-being of the poor and marginalized with access to crucial resources for survival. Social support and democratic processes were imagined as twin saviors of the developing world. Overall, the UN remained committed to a light brown ethos, putting social equity and justice first.

According to Ruether, the Bretton Woods institutions were major engines for development: the World Bank (WB), the International Monetary Fund (IMF), and the GATT, the General Agreement on Tariffs and Trade (which later became the World Trade Organization (WTO).[16] Together these bodies were responsible for the disassembly and reconstruction of many nations' economies, and what became known as Structural Adjustment. Typically, this meant moving underdeveloped nations toward development by requiring them to specialize in one or two commodities that could be sold in the global marketplace. Although the transformation was painful in the short term, the ultimate outcome, it was imagined, would be a world where the market took care of social inequities and intergroup conflicts. The Bretton Woods institutions promoted investment in underdeveloped nations, as Western corporations took the lead in the industrial extraction of resources (both natural and human) for their enrichment, and ostensibly for the greater good (Ruether 2005).

Meanwhile, development often disenfranchised and displaced the already poor and marginalized populations while enriching corrupt governmental and civil society actors who allocated aid with preference to the corporations who might line their own pockets. It also ignited in many colonized populations a resistance to colonial rule in general, and a reaction against capitalism and democracy specifically, which were seen as erasing traditional customs and lifeways. Distrust in Western hegemony proceeded in step with the spread of democracy and capitalism. Ruether noted that and attempted to expose the 500 years of imperialism that led to this impasse between West and East, or between the global North and the global South.

Under the rule of dictator Sadaam Hussein, in 1990 Iraq invaded its tiny Persian Gulf neighbor Kuwait. The Pentagon estimated that with that move, Hussein extended his control over about 20 percent of the global oil supply. Western powers reacted swiftly, condemning the action, and calling for Iraq's withdrawal. In league, George Bush, Sr., then President of the U.S., and Mikhail Gorbachev, the President of the Soviet Union, released a letter warning Hussein and Iraq to withdraw immediately. Margaret Thatcher, the UK's Prime Minister, added her vociferous condemnation, focusing international pressure. The UN released a resolution calling for immediate withdrawal, and shortly after, in 1991, the U.S. Congress approved military intervention on behalf of Kuwait, in Operation Desert Storm. The war ended swiftly, although Sadaam Hussein was neither captured nor removed from power at that time. It was no secret that the security interest of the US and European countries was oil. In March of 1991 Bush, Sr. gave the State of the Union Address presenting a vision of free, liberty-loving people engaged in a global war against Islamic extremism. Bush, Sr. opined: "A hundred generations have searched for this elusive path to peace, while a thousand wars raged across the span of human endeavor. Today that new world is struggling to be born, a world quite different from the one we've known. A world where the rule of law supplants the rule of the jungle."[17]

Just after Desert Storm ended Bush, Sr. gave the State of the Union address, offering what he called a "new world order":

> Now, we can see a new world coming into view. A world in which there is the very real prospect of a new world order. In the words of Winston Churchill, a 'world order' in which 'the principles of justice and fair play ... protect the weak against the strong ...' A world where the United Nations, freed from cold war stalemate, is poised to fulfil the historic vision of its founders. A world in which freedom and respect for human rights find a home among all nations.[18]

It is almost impossible to miss the faint waft of manifest destiny disguised as the globalization of individual liberty, free markets, and democratic ideals. Ruether noted the profound misunderstanding that guided U.S. foreign policy—that free individuals governed by democratic politics were themselves inevitably good. As the ships of freedom and democracy sailed across the globe, many social groups found this globalizing, leveling, commodified world a threat to their lifeways. Ruether's contribution shone a light on it. As she put it, "Deeper anti-Americanism surfaced among those who aspired to 'national liberation' from American-led neocolonialism. But efforts to shake free of this power and to foster alternative paths to development were undermined and defeated by a combination of economic strangulation through world financial institutions, embargo by the US, and either direct or surrogate military intervention."[19]

This dance between hegemonic powers and weak political actors who sought to disrupt Western imperialism came to a climax a decade later, on 11 September 2001, in the coordinated attacks on the World Trade Center, the Pentagon, and one other target which was not reached because passengers attacked their hijackers and brought the plane down. The reaction to the attack was international in scope as George Bush, Jr., the President at that time, portrayed the terrorist act as the latest battle in an ongoing and global war against Islamic extremists. In general, asymmetric warfare of weak political actors against global powers fostered imagined kinships that went beyond familial or political boundaries. Think here of the thousands that flocked to the *mujahadeen,* guerillas who fought the Soviets in Afghanistan, or to those Muslims worldwide who see the land conflict between Israel and Palestine as a common cause. In the last 70 years, all told, the weaker actors have won more such contests than they have lost. The Soviet Union attacked Afghanistan and withdrew after a decade of warfare. Later, the U.S. invaded Afghanistan to remove the Taliban, an ultra-conservative Islamic sect that was depicted as opposing freedom and democracy. After 20 years of fighting the U.S. also withdrew, and the Taliban seized power again almost immediately. A White House briefing described it thus: "Ultimately, after more than twenty years, more than $2 trillion dollars, and standing up an Afghan army of 300,000 soldiers, the speed and ease with which the Taliban took control of Afghanistan suggests that there was no scenario—except a permanent and significantly expanded U.S. military presence—that would have changed the trajectory."[20]

Ruether, keenly aware of the irony of American imperialism, sought to connect it to much older patterns of male-dominated dualisms which, given the added impetus of capitalism, transformed manifest destiny into a 20th century crusade to impose freedom on the rest of the world. Her work *America and Amerikkka: Elect Nation and Imperial Violence* (2014) traced the roots of American imperialism to some European attitudes which shaped America's vision of itself as exceptional. Ruether illuminated the ways in which claims of an "elect" nation or people always depended on discrimination against cultural, ethnic, or ethical others. Perhaps the most provocative suggestion was her recommendation that after the fall of American imperialism (which she located in the 2010s) a new vision of "letting go" might guide the liberation theology of the U.S.

Ruether's prescience, when it came to liberation theology, feminism, and environmental religions, was staggering considering many of her 20th century predictions materialized in the twenty-first century. The forcefulness of her writing welled up from her deep convictions about historical injustices and the asymmetrical exercise of power. Her contributions to the intersection of religions and environments were profound. Illustrating her contributions to a range of other scholarly avenues was a *festschrift* volume compiled and edited by a few of her most dedicated students on the occasion of her seventy-fifth birthday (Silverman et al. 2012). The essays ranged from studies of gender and public theology to extensions of Ruether's work on anti-Semitism and the spread of empire, and included voices who spoke of her influence across the globe. Ruether's legacy at the intersection of religions and environments included building ecofeminism, critiquing extractive capitalism, challenging the US's celebratory self-appraisal, and advocating instead a culture of peace and belonging. Ruther approvingly noted that one of her students highlighted three commitments derived from Ruether's teaching:

> 1) naming sexism and patriarchy as a perversion of human creation with which Christian theology and practice have historically colluded, which Christian practice is called to actively dismantle for the sake of all people and creation; 2) freeing practitioners of Christian theology to reclaim submerged portions of history and tradition in the service of dismantling structures of domination and creating more just communities and institutions; and 3) creating zones of safety for acknowledging and celebrating the individual voice and personal life story of the under-recognized as a meaningful starting point for theological reflection and action in the world.
>
> *(Silverman et al. 2012: xvi)*

That about sums it up.

QUESTIONS FOR DISCUSSION:

- Discuss the contours of Ruether's liberation theology. What was her solution to the patriarchal, dominion-oriented theologies of her time?
- Ruether was one of the first to expound a Christian ecofeminism. How did her vision relate to globalization? What was the purpose of a planetary ethos?
- What was the relationship between corporate globalization, and her prescriptions for challenging empire?
- What were Ruether's prescriptions for fighting corporate globalization and challenging empire?

Notes

1 The messianic hope was a phrase important to Judaism and to early Christian communities, and referred to the expectation that deliverance through a divinely anointed leader was imminent.
2 Some readers will recognize resonance with the work of the sociologist Max Weber's *The Protestant Ethic and the Spirit of Capitalism* (1920 [1905]; see Chapter 2).
3 Weber, however, was focused specifically on Protestant Churches.
4 See for example *Divini Redemptoris*, Pope Pius XI, 19 March 1937; *Laborem Exercens*, Pope John Paul II 14 September 1981; Evangelum Vitae, Pope John Paul II, 25 March 1995. See https://www.papalencyclicals.net/, accessed 6 September 2023.
5 The New Left referred to a growing movement in the 1960 and 1970s which advocated various social causes, from disenfranchisement of the poor and civil rights, to advocacy for women and gay rights.
6 Apocolypticism was the idea that a huge upheaval was coming, an "end times" when a new global order would emerge. It is widely accepted that the person of Jesus was an apocalyptic preacher, like many other itinerant messengers in the Near East at that time. Ruether's invocation of Platonism, concepts derived from the philosopher Plato, referred to the strong emergence of patriarchy and domination that depended upon a fundamental dualism between humans and nature. What was later termed neo-Platonism was very influential among early Jewish and Christian thinkers.
7 Interpersonal domination is the exercise of power over other individuals. Instances of racism or sexism, for example, are evidence of interpersonal domination. When such forms of domination become concretized in political, social, and/or economic realities they become systemic. The segregation of people in the U.S. based on race was an example of systemic domination.
8 *Oikos* is the Greek term for household. It is the etymological root of both ecology and economy.
9 Much of her work was influenced by Marx, and it is possible to see glimpses his thought in Ruether's writing.
10 The astrophysicist Brian Swimme was one of those who appeared in the acknowledgements. He popularized what he called "the Universe Story," and in cooperation with the Forum on Religion and Ecology, narrated a film collaboration called *Journey of the Universe*. See also the work of Michael Dowd, an evangelist for evolutionary Christianity.
11 The Abrahamic traditions are so called because they all trace their founding back to the person of Abraham.
12 For most scholars of religions the use of Judeo-Christian as an analytical term is passe, a relic from a Protestant construction of Christian history.
13 Masuzawa, for instance, wrote about the trajectory from Mohommedism (a regional cult) to global Islam (a world religion).
14 In addition to the five listed in Assisi, they added Confucianism, Daoism, Indigenous Religions, and Jainism.
15 See https://www.gtu.edu/news-events/events/lecture-address/other/teaching-for-peace-in-a-time-of-impending-war-2, accessed 8 August 2023.
16 The institutions were created at a meeting of Allied leaders in Bretton Woods, Maine (U.S.).
17 See https://millercenter.org/statecraftmovie/gulf-war, accessed 8 August 2023.
18 See https://www.c-span.org/video/?525425-10/a-world-order, accessed 8 August 2023.

19 See http://rl911truth.org/index.php/related-911-articles/54-ruether-rosemary-radford-american-empire-and-the-war-against-evil, accessed 8 August 2023. It can also be found in print in the fifth chapter of Kevin Barrett, John Cobb Jr., and Sandra Lubarsky, eds., *9/11 and American Empire: Christians, Jews, and Muslims Speak Out* (Olive Branch Press, 2007).

20 See https://www.whitehouse.gov/wp-content/uploads/2023/04/US-Withdrawal-from-Afghanistan.pdf, accessed 8 August 2023.

5

VINE DELORIA

"Sacred Geographies of American Indians: Negotiations and Networks" (American Indian Religions and Law)

For traditional Lakota practitioners and those trying to resuscitate the language and rituals, the Black Hills were the center of all things. It was from the Black Hills that the first humans emerged from the Earth. And for a long time, it was to the Black Hills that youths went to find spiritual orientation and renewal. In some ways, contestations over the Black Hills were a microcosm of the plight of Native populations in the Americas. The Sioux nation (a broader appellation which included the Lakota) was granted the Black Hills in the Treaty of Fort Laramie (1868). But once gold was discovered in the hills, the Lakota endured an illegal, forced removal from their ancestral home. Subsequently, from 1927 to 1941, the visages of four (white) U.S. Presidents were dynamited and carved into the face of the Six Grandfathers Mountain. The name was changed to Mount Rushmore following the public sculpting of sacred rock. It was for some Native peoples a slap in the face, their ancestral emergence place desecrated.

Vine Deloria (1933–2005) was born on the Standing Rock Sioux Reservation (Lakota), not too far away from the Black Hills, and he was one crucially important vector for the dissemination of American Indian scholarship in the United States in the late 20th and early 21st century. Deloria was a central figure in exposing the American public to American Indian religions and lifeways. He articulated the legal obstacles and pitfalls encountered when American Indians spoke about their religions to an ostensibly secular government and called attention to the public land disputes between American Indians and extractive industries. His work raised issues such as historical claims of sovereignty, the uneven application of laws related to religious freedom, and insights into how Indigenous knowledge might shape policy recommendations going forward, especially in tribal territories. Deloria was

DOI: 10.4324/9781003364801-6

situated at a pivotal time and place for American politics, Native sovereignty movements, and direct political action.

American Myths: Indians and the Popular Imagination

When American Indian spirituality exploded into the popular imagination in the United States in the latter half of the twentieth century it did so in a context where there was also a growing awareness of environmental issues. Nature, then, was from the beginning an important category for Deloria's work. Basically, Deloria argued that the Western preference for history over place was inevitably ecologically destructive (Martinez 2019). For American Indians, religious complexes were heavily dependent on specific geographic locales. For the Abrahamic religions in particular (Judaism, Christianity, and Islam) the emphasis was on a divinity that acted *in history* rather than in places. For instance, Christians believe that attending services in different cities or even countries still "works"—the human propitiation of that deity does not depend on a particular geographical location. In this, Deloria agreed with Lynn White, Jr.'s (1967) famous thesis that the drive for progress was embedded in the Judeo-Christian[1] dominion theme in Genesis. Deloria went further, suggesting that American Indian perspectives, which privileged the spatial over the temporal, were better explanations of reality than Western scientific models (Deloria 1997, 2003, 2006).

Deloria's writing was important because it was the first *American Indian* perspective to penetrate the academic world of religious and legal studies. Deloria challenged the domination of Native peoples, suggesting that the conflict between Christian history and Native practice was the primary cause of both social and environmental ills. During much of the 20th century, many non-Native people were oblivious to the existence of tribal peoples living on marginal lands.[2] During the latter half of 20th century, however, a series of potent public image events brought American Indians into the public eye. American Indians and their allies occupied Alcatraz Island in the San Francisco Bay, and stayed for over a year (Deloria 1992 [1973]: 9), calling themselves the "Indians of All Tribes."

Wounded Knee was the site where the U.S. Army massacred approximately 300 Lakota in 1890. The official reason for the Army's presence there was to quash an outlawed millenarian movement called the Ghost Shirt Dance.[3] In 1973 there was a direct action at Wounded Knee to renew American Indian sovereignty. This and other public image events elicited sympathy from many, and outright support from some of the nation's most well-known personalities (Deloria 1973: 9–10).[4] Deloria noted that during the 1960s many Americans were looking toward Indigenous religions, and Eastern philosophies and traditions for alternative sources of spiritual authority. And American Indian spirituality became one of the options in the marketplace of increasingly individualistic spiritual *bricolage*—the cobbling together of different concepts, practices for a personal spirituality (Deloria 1992 [1973]: 37).

Popular books such as Dee Brown's *Bury My Heart at Wounded Knee* (1971) and John Niehardt's *Black Elk Speaks* (1979) also fueled popular affinity for Indians. Interestingly, Niehardt downplayed somewhat the Christian elements of Black Elk's spiritual message, but Black Elk's Catholic commitments were exposed to the public, nonetheless. According to Deloria, Niehardt's work, along with Joseph Epes Brown's *The Sacred Pipe: Black Elk's Account of the Seven Rites of the Oglala Sioux* (1953), "formed a kind of sacred national Indian religious canon" (Deloira 1992 [1973]): 36). Frank Fools Crow, a traditionalist protégé of Black Elk who explicitly rejected Christianity in favor of traditional religion, led a group of Indians who, in 1975, had a hearing at the White House to advocate for Indigenous sovereignty. In the latter part of the twentieth century, American Indian religion was getting a hearing (although not a fair one by any stretch). But public appreciation of Indian culture harbored the potential for further cultural erosion from interested White people who wanted to help, but sometimes opened the door for more outside intrusion into tribal life.

The popular reification of American Indians stemmed, for Deloria (and incidentally, his son Phillip Deloria also (P. Deloria 1998)), from a desperate search for some genuineness in the marketplace of superficial values that gripped white America (Deloria 1992: 23). Vine Deloria stated that:

> the collapse of the Civil Rights movement, the concern with Vietnam and the war, the escape to drugs, the rise of power movements, and the return to Mother Earth can all be understood as desperate efforts of groups of people to flee abstract articulation of belief and superficial values and find authenticity wherever it could be found.
>
> *(Deloria 1992 [1973]: 52–53)*

Some Indigenous people like Sun Bear[5] sold their cultural heritage for profit, leading white seekers on vision quests, granting them Indian names, and not infrequently, seducing outsider participants (Deloria 1992: 40–41). Between Native and well-meaning White people there was exploitation in both directions. The romanticized caricature of the spiritually attuned Indian during the 1960s and 1970s played out in at least two ways. On the one hand, American Indians were saddled with the "noble savage" trope—not only were Indians spiritually superior, but they were also instinctual ecologists. American Indians were sometimes portrayed as savages who eked out a living in the howling wilderness. Or, just as damaging for their populations, they were imagined only in the past tense, as vanished cultures. Scholars such as Deloria and the scholar of religions and law Jace Weaver (Cherokee) rejected the notion that all Native peoples were inherently ecologically benign. As Weaver said, "in reality, modern Natives and their ancestors are neither saints nor sinners in environmental matters. They are human beings" (Weaver 1996: 7).

Anthropologists who focused on the connections between Indigenous populations and ecologies included the Swedish ethnologist Ake Hulkrantz (1981; Burhenn, 1997), who argued that the environment exerted a direct causal influence on the formation of religious rituals among Native populations. Their beliefs and practices were a result of the environmental constraints of their bioregion. Hulkrantz's observations were criticized for positing a mono-directional causalality, where the environments shaped human religiosity (Burhenn 1997). A more adequate theory, said his detractors, must also account for the influence of humans on their environment. The "world-fixing" rituals of the Plains Indians became a popular representation of Indigenous North Americans (Harrod 1995, 1993; cf. Charger 2013; Hand 1998). For the Plains tribes ceremonies such as the Lakota Sun Dance were directly responsible for the perpetuation of the world, and the cosmos. The function of such ceremonies was to negotiate exchange relationships between human and nonhuman worlds. The contrasting knowledge practices embodied by White Euro-Americans and American Indians grew into a feature of academic scholarship (Nelson 1983; de Castro 1998; Deloria 1985; Cajete 2015). For instance, the anthropologist Richard Nelson noted that the Koyukon people literally believed that the entire world, even inanimate entities, were alive, inspirited (Nelson 1983).[6] The world must therefore be negotiated with a great deal of respect for these other sentient entities. Nelson argued that their harsh environment necessitated strict codes regarding relationships with nonhuman animals and plants. Moreover, environmental fluctuations made certain resources unreliable. The Koyukon worldview, ingrained in cultural memory and carried on in religious ritual and oral tradition, reminded the Koyukon that they existed in a world of limits, a world that was animated by a thoroughgoing vitalism.

For Deloria, however, American Indian and Western European-derived cultural systems were incompatible, and White America had not learned to respect Native culture. Instead, they sought to usurp it. They represented two incompatible anthropologies and cosmologies. Anthropologies attempted to answer questions such as, "Why are humans here?" or "What is their purpose?" Cosmologies attempted to explain the existence and persistence of the universe. Western science depicted the unfolding of the cosmos as a linear sequence starting with a big bang. American Indian cosmologies were often viewed in terms of cycles rather than in linear terms—no big bangs.[7] But Deloria challenged those two prevailing myths about American Indians, that they disappeared, and that they were the first ecologists (Deloria 1973, 1975, 1985, 1999).

Escaping, Adopting, and Adapting Stereotypes

As the colonial enterprise gained steam in the Americas, many of the Indigenous populations were removed or destroyed, either by disease, or warfare, or outright murder. The disruption of ecosystems and the removal of people from their lands also disrupted social cohesion and the values systems of those peoples (see Hill 1996). Because Native rituals were deeply connected to the land, and because they began with a different understanding of the relationship between human and nonhuman worlds, they had a different epistemology, or knowledge-production system, so translation was tricky. The writer and ethnobotanist Gary Paul Nabhan (1987) referred to the disjuncture of perspective as "cultural parallax," which called attention to the difficulty in understanding cultural outsiders from inside our own worldviews and lifeways.

American Indian peoples lived with, and *did* alter their ecosystems significantly, making extensive use of fire and other tools for environmental modification. Nabhan's work challenged the assumption that American Indians were "first ecologists." Likewise, the anthropologist Shepard Kretch cautioned against oversimplified caricatures of Indian lifeways. Krech argued, echoing Nabhan that Natives altered their environment with fire, and maintained important relationships with several nonhuman animals (buffalo, deer, beaver, and more). Of course, Krech stated, it should be obvious that at first contacts no Natives would be aware of ecology in the Western, scientific sense of the word (indeed the term ecology only came into wide use in the twentieth century). American Indians were active agents in, not passive observers of, their environments, and they adopted functional behaviors to remain within their boundaries. Their imagined status as first ecologists likely derived in part from modern romanticization and the explosion of New Age interest in Indigenous spiritualities in the latter half of the twentieth century. Deloria, however, considered Krech's work an act of academic chicanery, and said Krech's conclusions derived from idiosyncratic and incomplete data (Deloria 2000). In part, Deloria was reacting to the scholarly "privilege" of a White anthropologist treating research subjects as data rather than embodied human beings who have long-standing traditions which underpinned their lifeways. Deloria believed the characters in Krech's book were without context. Native American symbols and tropes are, as are all symbols which carry authority, encountered, engineered, and re-deployed for different ends. Dismissing the environmental wisdom of American Indians by noting that they also sometimes overexploited their resource base, Deloria feared, was just another way to disparage the Native populations. Moreover, casting prehistoric aboriginals as the main villains of the mega- and meso-extinctions across the Americas did not help their public image, either.

The scholar of religions Sam Gill investigated widespread assumptions about American Indian spiritualities and values in his book *Mother Earth* (1989). Gill tracked the concept of the Earth as "mother," arguing that while many American Indians acknowledged the female energy of the Earth, it was Western scholars (beginning with Tylor, see Chapter 1) and the popular imagination that created the goddess-like "Mother Earth." Gill also discovered that the Earth was used as a metaphorical mother in more ancient American Indian contexts. The Mother Earth trope was also later adopted and adapted by many pan-Indian resistance movements, as well as in specific land battles. As Gill illustrated, American Indians have used and will continue to use such concepts for their own needs. They often adopt them when convenient (like the Mother Earth trope) and adapt them for their purposes. For example, Russell Means, one of the founders of the American Indian Movement (AIM, a radical Native activist group), used Mother Earth imagery to rally people for the protection of Native land and rights. Mother Earth also battled developers at Bear Butte, South Dakota, and recreationists at Mato Tipi (Devil's Tower) in Wyoming. While Mother Earth may have been a genuinely meaningful metaphor to some, it was also foregrounded when it was politically expedient to do so. The professor of law and religions Jace Weaver's *Defending Mother Earth* (1990) invoked this contested image and examined social justice issues in Native America. The view from where E.B. Tylor and James Frazer sat was that Native cultures were survivals from primitive cultures, unable to develop properly. But like all cultures, American Indians were and are dynamic, always adapting to shifting political circumstances and discourses.

One source for the notion that Indigenous people lived only in the past was the European preoccupation with the supposed Edenic "wildness" of the New World. Europeans and their descendants were led to believe that wilderness must be cleansed of human impacts before it could fit the bill of an American paradise. Thus, the "dispossession of the wilderness" spread across the continent (Spence 1999). The land was sanitized following its supposed neglect by Native populations, then fenced off for wilderness recreation. The idea of wilderness and the violence required to create state and federal parks popularized the view that nature was only truly nature when untarnished by human hands. That logic depended upon a dualism between humans and nature that most American Indians did not accept (see Chapter 8 for more on problematic understandings of wilderness).

Deloria, Weaver, and other scholars argued that we should do a better job of allowing American Indians to tell their *own* stories. Ultimately, what they want, as one of Shepard Krech's informants noted, is to be "just a real human being" (228). That paralleled Jace Weaver's insistence that modern Natives were "neither saints nor sinners," rather, they were "just people" (Weaver 1996: 7).

What was increasingly important was that the scholars who studied Native cultures grew in their awareness of just what it meant for Indigenous peoples to be real human beings. Recall the differing epistemologies of Indigenous peoples—they learned their stories by listening to the narratives of the land—literally being stalked by the places they inhabited (Basso 1996). Scholars were pushed to listen to Native populations in a different way, incorporating other peoples' science into land use and conservation behaviors (*cf.* Lansing 1991; Berkes et al. 1998; Berkes et al. 2003; Hukkinen 2008; Berkes 2012;).

Explorations of supposedly savage systems in South America resulted in a two-way exchange of sustenance practices, spiritual ideas, and unfortunately, diseases. Native peoples bargained with other entities in the nonhuman and preternatural worlds to preserve their own existence.[8] Historically shamans were responsible for maintaining right relations through spiritual brokerage. They interacted with spirits and animals to ensure the animals' return, thus ensuring the survival of the people. It is always dangerous to make overbroad generalizations, but perhaps it is safe to say that Native peoples North and South had in common a concern with reciprocity between human and nonhuman worlds. The exchanges were ritualized. Thus, religious ritual derived from, and was integrally related to, nonhuman nature. Religion, in such a system, acted as a homeostatic mechanism, keeping society within the bounds of its limited resource base (Rappaport 1966).

The anthropologist Gerardo Reichel-Dolmatoff was one of the early exponents of *ethnoecology*, the study of Indigenous taxa and ecological knowledge. In his extensive studies of Tukano-speaking peoples of South America (particularly Indigenes in the Northwestern portion of the Amazon), he detailed the relationships between cosmology and environmental behavior. The plants and animals that comprised the environment of the tribes were literally conceived of as biological extensions of the people. So, for them a distinction between humans and nature was mistaken (the Desana, e.g., have no word for nature). The Earth was not a Gaia-type entity with the ability to self-regulate. Instead, the Desana (and other neighboring tribes) had a responsibility to minister to the world to ensure its continued survival. Without their rituals, the world would collapse. The energy flows that perpetuated the world were coded into religious rituals. Other tribes performed rituals with parallel aims, ultimately taking on themselves the responsibility for the perpetuation of the cosmos (Crocker 1985; Wright 1988; Harrod 1995). The colonial enterprise resulted in contested cosmologies and practices (see, e.g., Taussig 1978, 1980, 1987; Wright 1988; Whitehead and Wright 2004; Vilaça & Wright 2009).

Natives North and South: Residence and Resistance

The timeline for the residence of Indigenous peoples in the Americas continues to be pushed back deeper into the past. Archaeologists and anthropologists developed ever-better remote imaging techniques that exposed settlement patterns even where there was no observable presence of cultural artifacts. When I learned about natural history as a child, the prevailing wisdom was that American Indians emigrated to the Americas by passing over the Bering Land Bridge. This bridge between North America and Northeastern Asia was exposed during the last Ice Age when much of the planet's water was taken up into ice flows, approximately 37,000 years ago. Paleolithic hunter-gatherers supposedly ventured across following migrating food sources approximately 16,000 years before present. Although that land bridge was likely used by social groups to move into the Americas, scientists now know that humans were also here long before the bridge appeared. At this writing, another spectacular find received wide attention—a set of footprints pressed into the paleolithic soils of what is now New Mexico which date to 21,000–23,000 years old.[9] At the Monte Verde site in Southern Chile, evidence placed humans at the far Southern end of South America by approximately 18,000 BCE. Native occupation of the Americas is, then, more ancient than previously imagined. This unfolding story of human appearance in the Americas raised larger debates about how long humans occupied these continents and how they related to their changing habitats over time.

For some Native populations the performance of many Indigenous religious complexes today depends on preserving their historical authenticity. Connecting current practices with past traditions, lineages, and events is one way to add some credibility, and some authority to contemporary cultural practices and ethical obligations. Some Natives perceived that the burgeoning interest in Indigenous causes across the globe in the late twentieth century should be leveraged to draw more people of European heritage into an understanding of Native lifeways. In any case, historically, environmental factors shaped religious complexes. And in many cases, religious rituals and practices modulated human impacts on the environment.

Most written data regarding pre-contact peoples came from post-contact European writings. Cabeza de Vaca, for example, kept memoirs of his *Adventures in the Uncharted Interior of America* (Cabeza de Vaca 2003 [1961]) in the early sixteenth century, providing a relatively favorable description of the Natives he encountered. De Vaca and his crew were shipwrecked on the coast of Florida, but continued their intrepid exploration on hastily assembled barges, eventually landing on the Eastern coast of modern-day Mexico. De Vaca traded with, and kept records of the people he met, eventually "going native" himself—supposedly living naked, off the land, with a retinue of Natives. The sixteenth-century monk Bartolome de las Casas was perceived by some as an advocate for Native peoples. Many interpreted his accounts as evidence of a positive relationship between settler intruders and Native peoples. But when it came to the treatment of Natives, he was part of a patronizing relationship of domination of the Indigenous people of Hispaniola. Although he perceived himself to be a protector, the shameful legacies of those who came after exposed a poisonous patronage which resulted in the torture, enslavement, and decimation of Indigenous populations from the earliest encounters.

When the European encountered the Maya, they were a large and advanced civilization in Meso-America. The Mayan culture dated back to the second millennium BCE and reached its peak between the seventh and ninth centuries CE. Local populations and resources were drafted into service in competitions between local chiefs intent on displaying their wealth and power (Diamond 2011). The preoccupation of those powerful elites with material aggrandizement caused them to turn a blind eye to growing problems with deforestation and drought. Although the Maya created an extensive trade network including several city centers, fluctuation in environmental conditions radically destabilized their ability to continue traditional modes of production and subsistence.

In the high deserts of North America, the ancestral Puebloans (pejoratively called the Anasazi) thrived for over a thousand years. At its height, the Chaco culture encompassed several towns and stretched over several hundred miles.[10] The center of the culture was Chaco Canyon itself, with at least four large structures, some of which were ceremonial centers, while others acted as residential units. The largest, Pueblo Bonito, had over 800 rooms. The character of such settlement suggests that Chaco may have been a sort of hub for the outlying towns. Although the large number of rooms suggest massive storage areas for grain, maize, and other crops, evidently little food production occurred in Chaco Canyon itself. Around 1150–1200 CE, those peoples promptly vanished from the high desert, leaving their dwellings to tease the imaginations of generations to come. One of the ways in which archaeologists traced these developments was midden dating, examining piles of dung to reveal secrets about their lives. Such evidence allowed scientists to more accurately reconstruct the conditions that led to the decline of the proto-Puebloans. It became increasingly clear that both deforestation and drought on the high mesas by the middle 1100s forced groups to abandon their settlements. They migrated mostly east and south, founding a series of Pueblos, many of which remain today (Zia, Taos, and others in New Mexico and Northern Mexico). Diamond (2011) suggested that the proto-Puebloans were unaware, as were the Maya, of the narrow environmental range that could continue to support large human populations. Both cultures were centered on a group of elites who demanded large resource tributes, and who did not, themselves, produce any goods or services.

In some cases, then, pre-Columbian peoples were no better than any other humans at sustaining their resource base. As we have seen many anthropologists (i.e., Shepard Krech) argued that Indigenous populations sometimes participated in extensive alteration of the land, as well as plant and animal regimes during the entirety of their tenure in the Americas (see Nabhan 1997; Krech 1999, 2005). It mattered little in the end. The differential social development of Europe and the Americas meant that American Indians had no immunity to the passel of bacterial and viral diseases transferred from animal to human and back again in the close urban centers of Europe (Eisenberg 1998). The historian Alfred Crosby (2000) argued that the environmental circumstances present in Europe at the beginning of the age of exploration ultimately determined the fate of the American Indians. Europeans had long coexisted in close proximity to domesticated animals before they came to the Americas. Europeans had in some cases developed immunities to illnesses that decimated Native popoulations. It would not have been long before the continent looked, as many Europeans imagined, like a massive, uninhabited resource bank. Europeans (both intentionally and unintentionally) imported invasive plants and animals to the New World, transforming the Americas into a vassal state of the European monarchies. Crosby termed this transition "ecological imperialism" (Crosby 2000).

According to the historian and bioregional[11] advocate Kirkpatrick Sale, there were three main trends that arose from the Enlightenment and contributed to the conquest of the American paradise: humanism, rationalism, and materialism (Sale 1990). To elaborate, (1) Humanism included the belief in the excellence and sublimity of the human being; (2) rationalism meant that all knowledge was reducible to empirical statements; and (3) materialism was a preoccupation with material wealth and power. These became the engines of modernity. For Sale, those trends were created and reinforced by the miserable state of Europe around the time of Columbus's journeys. The Black Death was still making the regular rounds. Some major cities experienced over ten outbreaks within 200 years, with a cost of about half the population of Europe. Famine was also a regular occurrence, and according to Sale, violence, perhaps owing to the more crowded urban conditions, skyrocketed. In the end, Sale suggested that the overriding preoccupation for Europeans was "how to survive the misery and suffering and violence that seemed to be rushing the world to its end" (Sale 1991: 46). The answer for many Europeans, as Christopher Columbus's little fleet headed due west from the Canaries, was the conquest of the American Paradise. The conquest was often imagined as the European destiny, obscuring the violence of colonialism with appeals to the Christian god. This was history through Deloria's lens.

The Law and the Profit: Deloria on Kinship and Religious Freedom

Pope Alexander VI was a particularly unsavory character in Deloria's eyes. His encyclical *Inter Catera* (1493) granted the "New World" (which of course was new only to Europeans) to the Spanish. After Portuguese protestations, a line was drawn giving Portugal a smaller slice of South America to do with as they wished. Ironically, of course, the land was not the Pope's to give. "What this pious language meant in practical terms," said Deloria, "was that if confiscation of lands were couched in quasi-religious sentiments, the nations of Europe could proceed" (Deloria 1973: 265; Glass 2005). Far-removed monarchs ceded lands that were not theirs to give to curry favor with religious leaders. The Europeans who came to the Americas dwelled in the mechanistic, disenchanted world described by Carolyn Merchant (see Chapter 6). One characterized by the persistence of dualism and the pursuit of objectivity which callously cast aside moral obligations to other creatures, and trumpeted an exclusivist religious stance. The acts that destroyed Native populations were precipitated by a different mode of production, a different superstructure, and thus a different way of being-in-the-world. The primary reason for this cultural incommensurability was the American preference for the historical over the natural—so it is worth digging a little deeper into this impasse (Deloria 1996: 58). While most Christians viewed revelation as an objective event that occurred in the past and remained applicable to modern life, revelation for Native peoples was "a continuous process of adjustment to the natural surroundings and not as a specific message valid for all times and places" (Deloria 1996: 67). Moreover, "the essence of the Indian attitude toward peoples, lands, and other life forms is one of kinship relations in which no element of life can go unattached from human society" (Deloria 1999: 131). Kinship derived not from blood-borne relationships, but rather through intentional encounters with other persons. The Indigenous world was populated with all kinds of persons, only some of whom were human (see the connections with animism in Chapter 1). Kinship relations were important features of the sense of etiquette that enveloped both human and non-human communities. These close ties required the performance of reciprocity and respect. As Deloria put it, "the essence of this behavior is the maintenance of dignity throughout the organic world" (Deloria 1999: 131, 145–155; *cf.* Cheney & Weston 1999; Harvey 2013).

Native populations, argued Deloria, maintained conscientious relationship with places, rather than identification with historical figures (cf. Weaver 1996: 13). As Deloria put it, "Western religions concentrate on the idea of history and what it has revealed to human beings about the nature of God. Both the Christian and Jewish religions are dependent on the recitation of historic events and their interpretation as the means of determining the validity of their beliefs" (Deloria 1999: 126). "Almost every tribal religion," he countered, "was based on land in the sense that the tribe felt that its lands were specifically given to it to use.... lands cannot be sold since no human has the power or right to own them" (Deloria 1999: 127).[12] If Deloria was right, and this intense attachment to the land is only possible through extended occupation of places, then perhaps the best solution is to grant Indigenous people sovereignty over their lands once again (Deloria 1996: 20). Only then can Native people drink deeply of the land, perhaps renewing lifeways that provide models for sustainable, and equitable life with the land.

If this sounds like a somewhat vague or idealistic solution, it is. Native groups are on average some of the poorest in the Americas. The U.S. legal system often dismissed Indigenous spiritual practices and beliefs as not sincerely held, or not evidencing sufficient structures (physical or ecclesial) to "count" as a religion. Although promised recognition and reconciliation, the Native experience was mass death and further marginalization. The Northwest Ordinance of 1787 made many important changes to the governance of the young United States. It established the Northwest Territory. It laid out the process for adding new states to the union. And the law promised that no unceded lands would be taken from the Indigenous inhabitants *except* in *just* wars. The Treaty of Ft. Laramie (1868) designated the Black Hills in the Dakotas as a home to Siouxan tribes in perpetuity. The promises did not last long. By 1876 U.S. cavalryman George Armstrong Custer acted against orders and took a party into Native territory following the news that gold was found in the Black Hills. Combined American Indian Forces (Lakota, Cheyenne, and Arapaho) annihilated Custer's unit near the Little Bighorn River. In 1877 the land was confiscated by the U.S. government with no agreement or even consultation with the Native tribes.[13] This was just the beginning of the trail of broken treaties (Deloria 1974).

Deloria was indignant toward those who tacitly endorsed American Indian genocide and dispossession by declaring that those people who murdered and cheated Indigenous Americans "back then" must not have been *authentic* Christians. Surely they had been misguided, cried their modern-day counterparts (Deloria 1999: 261–262). Deloria eviscerated such arguments illuminating the point that just as Christians of days past had considered themselves the embodiment of the Christian vocation, those who did so in the present also perceived themselves as an authentic manifestation of Christianity. Moreover, Deloria argued, if those exploiters of old were not Christians, then "why did not the true Christians rise up in defiance of the derogation of their religious heritage and faith?" (Deloria 1992: 262). Dismissing that line of argumentation, he wrote that "It is time to call a halt to the unchallenged assumptions of the Christian conception of history" (Deloria 1996: 262). "We cannot reject the Christian religion piecemeal," Deloria forcefully responded. "The importance of the decline of the specific doctrines of the Christian religion is that this decline means that the whole religion has been misdirected from its inception" (1999: 265). The argument paralleled that offered by the Civil Rights leader Martin Luther King, Jr. (1963) in his "Letter from a Birmingham Jail," (1963) in which King asked the well-meaning liberal White clergy and practitioners why they were not in the streets marching with their oppressed brothers. Both were powerful condemnations of sympathetic constituencies that looked the other way when they caught a glimpse of the gritty underbelly of colonial power. Facing that dark underbelly would require societies to confront their own naturalized presuppositions. In this context, something is *naturalized* when it becomes something that "goes without saying." For instance, among E.B. Tylor and his contemporaries, it went without saying that all non-Western people must be more primitive than were the Europeans.

Deloria, Weaver, and many collaborators and followers stared down Western histories which portrayed American Indians as living in the past, not the present, and certainly not into the future. What many Native populations wish for is recognition of their plight in the present, and measures to address the drought, unemployment, high rates of drug abuse, suicide, and crime that plague many reservations. In many cases, they need resources (from survival staples to educational materials) for use at their own discretion. Not resources tied to government oversight and strictures. As Weaver noted, what American Indians ask is to be treated as human beings—no more, no less. As with any American tribes their resource travails fail to receive the same sort of policy attention that other communities do. As movements across the globe deploy tropes and policy recommendations that characterize specific places as holy, sacred, they are capitalizing on the "ontologization" of the sacred to create the legal friction for potential change.[14] There are emerging consonances between the ways in which different traditional and tribal groups have utilized the sense of the sacred to delineate and protect lifeways. This may become an even more important tool for policy development going forward.

Historically, Native lands were stolen whenever there was money to be made from them. When gold was discovered in the Black Hills, the land was overrun. When it turned out that Indigenous populations occupied land with valuable resources, they were driven off it, or their rights to those resources were sold without their consent. In all these cases, the government helped to engineer the seizure of those lands by declaring Indigenous religions inauthentic. In some cases, the courts declared that they might not be *real* religions because there were no buildings or other constructions that demonstrated their sacredness. For instance, on Black Mesa, a massive coal mining operation caused dramatic losses to the Navajo peoples' aquifer. Many shrines on the Hopi reservation were destroyed because there were no human-made structures and because their practices forbid them from exposing their locations to outsiders. In other cases, the definition of religion was reduced to a "sincerely held belief," a definition which also denied American Indian religions authenticity. In the case *Employment Division v. Smith*, a practitioner in the Native American Church (NAC) was fired for imbibing peyote as part of a religious ritual. Peyote is a spineless cactus which contains mescaline, a powerful hallucinogen, and is a Schedule I drug in the United States. In this case, although people were free to *believe* that their religious practices required the use of peyote, they could not *practice* such rituals. When it came to American Indians, the courts have been intentionally antagonistic, and at best ambivalent. In no cases, however, were the religious complexes of American Indians considered to be widely accepted or protected.

QUESTIONS FOR DISCUSSION:

- Were American Indians "first ecologists," or "noble savages"? Describe what these phrases mean, and how they affected Native populations.
- Can you think of any examples where Indigenous populations have used public exposure to call attention to Indigenous land sovereignty?
- What is the importance of kinship in Deloria's work? Who is included in his descriptions of kin relations?

Notes

1 I use Judeo-Christian in this context because Deloria, Lynn White, Jr. (and many others) used this phrasing. It is typically no longer used as an analytical term as it implies some common history that is a product of Protestant millenarianism.

2 In most cases, Indigenous populations occupied meager habitats because they were dispossessed of their own land or had their footprint radically reduced through seizures by the federal government.

3 The practice of the Ghost Shirt Dance was spread by a Paiute named Wavoka (c. 1856–1932) as a form of resistance to settler colonialism. An earlier, similar movement attributed to another Paiute, Wodziwob, preceded Wavoka's evangelization. The dance, and the accompanying shirt were supposed to protect the wearer from White bullets. The goal of the Ghost Shirt Dances was to call the spirits of the dead, and to end theconflict between cultures, ultimately resulting in the return of American Indian homelands and traditional lifeways.

4 For example, the famous actor Marlon Brando did not accept his 1973 Oscar award for the film *The Godfather*, and instead a young Native woman, Sacheen Littlefeather, attended the Oscar ceremony and read aloud a letter penned by Brando, rejecting the award in protest of the treatment of American Indians.

5 Sun Bear was a "medicine man" of Chippewa descent who claimed he started a new tribe of his own, encouraging people to join and contribute money. Sun Bear wrote several books on spirituality, but he is an example of what came to be known as a "plastic medicine man"—that is, inauthentic. For more see Churchill (2010), https://www.culturalsurvival.org/publications/cultural-survival-quarterly/spiritual-hucksterismthe-rise-plastic-medicine-men, accessed 30 May 2023.
6 The Koyukon are a tribal group whose ancestral home is in Alaska.
7 It should be noted that at this writing the popular version of universal genesis in a Big Bang is not the only theory used by physicists to explain the beginning of our universe. During the latter half of the twentieth century, however, this was the most widely accepted theory.
8 The term preternatural refers to things that are out of the ordinary, or which somehow exceed their normal or natural state (not necessarily *super*-natural).
9 See, for instance, National Geographic's coverage (https://www.nationalgeographic.com/history/article/fossil-footprints-challenge-theory-when-people-first-arrived-americas), accessed 5 June 2023.
10 The center of the Chaco culture was Chaco canyon near the Four Corners area in present day New Mexico.
11 For more on bioregionalism as a mode of social organization, see the Chapter 8.
12 Social constructionism describes a position in which someone perceives that trying to find some *essence* that is considered sacred across cultures is wrongheaded. Constructionists perceive that the very use of language as a structured form of communication, and our perception of the world through our limited biological senses, is a representation which cannot escape the contingency of our perceptual abilities.
13 It was not officially a war, and thus could not be the cause of the legal removal of treaty rights.
14 See the Chapter 3 for more about a theoretical stance that treats the sacred as *really real*—or, ontologized.

6

CAROLYN MERCHANT

"The Death of Nature and an Ethic of Care" (Environmental History)

Ecofeminist Environmental History

After exploring the work of Rosemary Radford Ruether (see Chapter 4), it is fitting that we work our way toward the monumental environmental historian Carolyn Merchant (b. 1936). Careful readers will note the resonances between Ruether's and Merchant's versions of an ethic of care. Environmental History was not new, but Merchant's contributions were crucial interventions in that area. Some of the most important environmental histories prior to Merhant included Clarence Glacken's *Traces on the Rhodian Shore* (1967), Roderick Nash's *Wilderness and the American Mind* (1967), and Donald Worster's *Nature's Economy: A History of Ecological Ideas* (1977). I review their contributions here to provide context for Merchant's unique contributions.

Glacken's *Traces* plots the history of Western philosophical thought from the Ancient Greeks up to the death of Alexander in 323 BCE, and into the beginning of the Roman *Pax Romana* around 30BCE (when Augustus was Emperor). There was an epistemological and conceptual chasm between pre-Hellenic and Hellenic societies (Glacken 1967: 25). Fundamental to this distinction is the idea that the world is ordered, unified, and harmonious and that there is some reason or *purpose* in nature. Here are the first stirrings of teleological thinking (although it would not mature until Aristotle)—where the root *telos* means purpose or end. Contestations regarding the question of purpose would occupy Western thought for millennia afterward. In broad strokes, however, Glacken's analysis suggested that there was a longstanding tension in Western thought between organic metaphors, a sense of belonging, purpose, and harmony, and those which might be called mechanical,

DOI: 10.4324/9781003364801-7

conceptualizing the world through regimented, bureaucratic, and pragmatic eyes. Merchant's contributions focused on later time periods, but her work illustrates her understanding of these ancient questions.

Conservation in the United States grew from urban movements in Europe, spreading shortly thereafter to the more urban Northeastern sections of the United States. Similarly, late twentieth-century concepts of global conservation spread from the urban industrial centers of the United States and Europe to the developing world, including Asia and Africa. In both cases, the spread of environmental sentiment was almost always rendered in capitalistic economic terms. Historian Roderick Nash argued that "thinking of wild nature as an actively traded commodity in an international market clarifies appreciation and largely explains the world nature protection movement" (Nash 1967: 343). In a sense, Nash wanted his readers to recognize the irony of victory. Even though environmental consciousness became international in scope, it remained an ambivalent force, encouraging to some extent the commodification of nature, and sometimes compromising the lifeways of already marginalized Indigenous peoples. Moreover, Nash perceived, there was a positive feedback relationship between the vision of a howling wilderness, and the eschatological hopes of those who saw that mastering it was their religious vocation.

Daniel Worster's *Nature's Economy* (1979) traced the history of the science of ecology from its eighteenth-century roots to modern manifestations, highlighting the historical contingency of scientific theories and methodologies. Worster suggested that the science of ecology consistently generated two distinct interpretations of nature: the arcadian versus the imperial, or the organic versus the mechanical (Worster 1977: 2). The arcadian vision, epitomized in the naturalist Gilbert White, envisioned a vital force animating the cosmos, and a vision of peaceable co-existence with the non-human world. The imperial tradition, represented by Linnaeus and others, acted to order and dominate nature through the appropriate application of reason and hard work (Worster 1979: 2). Worster suggested that:

> fortified by the principle of historicism, we can approach recent ecological models that dramatize disturbance with a sense of skepticism and independence. If they are not the mere reflection of global capitalism and its ideology, they are nonetheless highly compatible with that force rearranging the earth. The newest ecology…is congruent with…'the logic of late capitalism'
> *(Worster 1979: 427)*

This was a harsh criticism of a science often equated with conservation aims (*cf.* Davison 2001). One key contribution that Worster made was to illustrate that the concept of an ecological system was much older than the science of ecology itself. He explored two competing and persistent interpretations of how nature "works." Much like Glacken (1967), he historicized scientific social formations, noting that the mechanistic view of that time was countered by a cyclical return to organicism. These were important additions to the religion and nature ferment.

Merchant went where those authors did not—she criticized the lack of women's voices, and expanded the canon of historical authorities to include women and other marginalized classes. In so doing she highlighted that contributions of women were often muffled by male-centric, scientific, political, and economic development. The book for which Merchant was best known was *The Death of Nature: Women, Ecology, and the Scientific Revolution* (1980), which was long a staple for those studying the nexus of religions, nature, and culture. In the book she traced changes in the intellectual milieu from roughly the 1500s through the 1700s, focusing on the sexual politics of the men (the gendered noun is intentional here) who generated and sustained the emerging scientific method. As the environmental philosopher Robyn Eckersley (1998) offered,

> Merchant's examination of changing scientific thought and practice is located within the context of a broader theory of cultural transformation that looks at the economic, ecological, social, religious, and philosophical context in which scientific transformations take place. This is not a detailed examination of the internal development of science. Rather, it is a broad investigation of the external context in which scientific ideas emerge, take root, flourish, and evolve.
>
> *(Eckersley 1998: 184)*

Economic and technical developments created a positive feedback loop which reinforced patriarchal political arrangements and emerging capitalist markets with a narrative that relegated women and other marginalized groups to subsidiary roles in history. Major themes in her work included the transition from matriarchal to patriarchal, from mechanical to organic, from nature as disordered matter to managed profitable material. Hierarchical societies with patriarchal religious groups dominated cultural, ethnic, and ethical others. The mechanistic perspective offered by science portrayed the world as dead inert matter, while the propitiatory behavior of cultural, ethnic, and ethical *others* were cast as primitive or savage explanations for things unknown.

Merchant drew on Ruether's earlier work, and they remained in close contact. Recall that Ruether's work focused on Catholic Christianity, one of many groups which imposed the transition from maternalistic, collaborative worldviews toward dualistic, combative, and mechanized worldviews. Perhaps Merchant's most important contribution for the study of religions and the environment was to expand the critical ecofeminist perspective to environmental history and to the humanities in general.

Production, Reproduction, and Revolutions

It is worth digging deeper into the historical arc of Merchant's critiques of modern science. In *The Death of Nature* (1980), Merchant began with transitions in the mode of production. What was once an organic whole embodied by the divine feminine, was bulldozed by a mode of production that aimed at mastery and command. The role of women shifted from providers who managed agricultural plots and gathered for family groups to a life that was primarily about tending to the *inside* of the home. The characters in this tale were mostly European. Feudalism eroded thanks to expanding trade networks and the growth of a new merchant class. The result was an increasingly aggressive use of more advanced agricultural technology. The transition was justified with reference to creator beings who had designed humans to be little less than angels, and certainly qualitatively more than the rest of creation (Merchant 1992: 52-53).

In *Ecological Revolutions: Nature, Gender, and Science in New England* (Merchant 1989), she cast her gaze to the political geography of the Northeastern United States. She recounted two revolutions whose roots emerged from the Northeast: the Colonial Ecological Revolution and the Capitalist Ecological Revolution. "My thesis," Merchant said, "is that ecological revolutions are major transformations in human relations with nonhuman nature. They arise from changes, tensions, and contradictions that develop between a society's mode of production and its ecology, and between its modes of production and reproduction" (Merchant 1989: 2). To make such an argument Merchant asserted that nature was a "historical actor" (Merchant 1989: 7). The ecological revolution took place between 1600 and 1675, while 1775–1860 was the setting for the capitalist revolution. While the colonial revolution would increasingly solidify the emerging mechanical view of the world, Merchant argued that some people and groups retained survivals of the earlier organic worldview. But "under capitalism, the division of labor between the sexes results in a situation in which men bear the responsibility for and dominate the production of exchange commodities, while women bear responsibility for reproduction of the work force and social relations" (Merchant 1989: 17). The rise of mechanistic interpretations of consciousness was a product of the capitalist revolution, which justified the identification of the female with nature, and resulted in the subsequent domination of these categories (*nature* and *female*). As she surveyed the transition from the colonialist to the capitalist revolution she observed that during the scientific revolution,

> The new definition of reality...was therefore consistent with, and analogous to, the structure of machines. Machines (1) are made up of parts, (2) give particulate information about the world, (3) are based on order and regularity (perform operations in an ordered sequence), (4) operated in a limited,

precisely defined domain of the total context, and (5) give us power over nature. In turn, the mechanical structure of reality (1) is made up of atomic parts, (2) consists of discrete information bits extracted from the world, (3) assumed to operate according to laws and rules (4) is based on context-free abstraction from the changing complex world of appearance, and (5) is defined so as to give us maximum capability for manipulation and control over nature.

(1989: 52–53)

Merchant weaved in themes present in other earlier works, highlighting the shift in land use patterns in New England and important variables such as the calories available to both humans and livestock. She kept in view Frederick Jackson Turner's famous "frontier hypothesis" (Turner 1920), which offered an analysis of the important role of the westward-moving frontier and manifest destiny in the moral anthropology of Americans. That vision of encounter and control transformed animals into resources, and moved the mode of production from subsistence farming and foraging toward commodity production. The land witnessed the flight of the "corn mothers" (a reference to the maternalistic personification of the forces that helped the corn grow) upon the arrival of the Puritan Fathers (who sought to domesticate the howling wilderness and enforce a theocratic society preoccupied with purity) (Merchant 1989: 69–111). Merchant described the reconstruction of subsistence farms into market farms: "The vitalist-mechanist debate in agriculture was a contest for power over the terms of scientific discourse.... In the process of reconstruction, muck, *geine*, Gaia, and vital forces—the animate world's terms of discourse—fell as the laboratory rose" (Merchant 1989: 211).

Merchant demonstrated how these solution-oriented sciences joined with a dualistic metaphysics to affect both the perceptions and functions of women. "The capitalist ecological revolution," Merchant argued,

"*split production and reproduction into two separate spheres*. As male farmers began to specialize in women's traditional dairy, poultry, and vegetable production, and as textile mills took over their clothing production, woman's primary domain was redefined as reproduction. Woman's sphere still included her biological and social roles of childbearing and childrearing...but her traditional farmyard and garden production was constricted to the preproduction of daily life within the household walls" (233, emphasis added).

Categorization and *naming*—tools of conquest—were handmaidens of the corpuscular, mechanized, and masculine worldview.

The scholar of religions Russell McCutcheon argued that classification and naming is always an exercise of power. He used as an example the mountain commonly known as Mount Everest. Does it matter that the person for whom it is named—Everest—was an Englishman who never set foot on the mountain? Or that he pronounced his name "Eve-rest" (rather than ever-rest)? What do the Indigenous peoples that live in the area gain when they deploy their name for the mountain, Sagarmatha? (McCutcheon 2019). Surveying, cataloging, and classifying during the colonial period and beyond modified the human geography by claiming spaces and material objects for Western monarchies. The first European botanical gardens and zoos displayed exotic flora and fauna from faraway places, extracting them from their home and keeping them for the amusement of Western onlookers. People, too, were trafficked in these colonial extractions (Chidester 2003). Those cultural others who bore the yoke of cultural imperialism became the trophies of the most dangerous game—industrializing humans. Merchant made clear the link between the mode of production, the rational gods'-eye view offered by science (objectivity), and the objectification of women in Europe and America. Specifically, women experienced increasing alienation from meaningful collaborative labor as their work was forced indoors. Merchant argued:

> A subsistence economy in which resources, goods, money, or labor were exchanged for commodities was replaced in many areas by the open-ended accumulation of profits in an international market. Living animate nature died, while dead inanimate money was endowed with life. Increasingly capital and the market assumed the organic attributes of growth, strength, activity, pregnancy, weakness, decay, and collapse, obscuring and mystifying the new underlying social relations of production and reproduction that made economic growth and progress possible. Nature, women, blacks, and wage laborers were set on a path toward a new status as 'natural' and as human resources for the modern world system. Perhaps the ultimate irony in these transformations was the new name given them: rationality.
>
> *(Merchant 1992: 56)*

Merchant always returned to her main themes and arguments, carrying the impactful analysis of *The Death of Nature* (1980) into all her subsequent work: "The removal of animistic, organic assumptions about the cosmos constituted the death of nature—the most far-reaching effect of the scientific revolution" (Merchant 1992: 47). *Radical Ecology* (Merchant 1992) provided a capsulized analysis of arguments from earlier works and placed the rise of rational scientific investigation in conversation with late 20th century social movements. One of the movements she analyzed was Deep Ecology, a philosophy of place offered by the Norwegian philosopher Arne Naess (1912–2009). Deep Ecology celebrated the intrinsic value of nature with human beings *in* it, not apart from it (Naess 1973). It relied on the celebration of diversity, localized solutions, and adaptive systems. This philosophical orientation was impactful in the field of religion and the environment, prompting responses from many environmental philosophers. It also provided ammunition for outgunned environmental activists. Radical environmentalists enacted grassroots campaigns to arrest ecological harms or to call attention to them through public image events. For instance, in the first direct action of Earth First!, perhaps the first radical environmental organization, they unfurled a very long, black sliver of plastic down the face of the Glen Canyon Dam—resembling a giant crack—to call attention to the landscape drowned by Lake Powell.[1] Although Merchant felt their passion was commendable, she averred that Deep Ecology was thoroughly embedded in the patriarchal landscape, reifying principles which to some seemed universalizing and insufficiently revolutionary. Merchant's interpretation of Deep Ecology was that "Naess' philosophy has only one ultimate principle—Self-realization….as one's own self-realization expands outward into the great Self. This process of increasing Self-realization was described by poet Robinson Jeffers as 'falling in love outward'" (95). Grounded in binary oppositions, reflected in mechanistic science, such ideas were birthed by patriarchal societies.

The science of ecology (a version that Naess would have termed the "shallow, short-range ecology movement"), Merchant felt, was precisely the sort of scientific development that might lead to a sustainable post-industrial society. She went on to invoke the Gaia Theory, pioneered by independent scholar James Lovelock and the microbiologist Lynn Margulis, which imagined the earth itself as a homeostatic entity that pursued its own sort of dynamic equilibrium state. Unlike mechanistic science, she argued, "ecology is based on open, rather than closed systems and emphasizes nature as continuous change and process" (105). Gaia, like ecology, projected a multi-scalar vision of Earth as a self-regulating super-organism, and was process-oriented and open. Merchant's equation of the concept of Gaia with the science of ecology was provocative. Hers is an interpretation of *ecology* that goes beyond what most ecological scientists would be comfortable with. Merchant's version is not just about the science of interactions between individuals, their populations, and their ecosystems, it is also a holistic way of perceiving the natural world. Merchant extended the holism beyond the ecological into the cosmological, influenced by the physicist Fritjof Capra (1975), who used Daoism to introduce the Western world to the physics and metaphysics of connection. Capra, inspired by ingesting "power plants" containing the entheogenic compound psilocybin, offered a view of the cosmos itself as inhabited by a pervasive and interactional force that we might, under the right circumstances, commune with and use to shape our perception. Similar arguments precipitated out of scientific communities thereafter. For instance, Merchant also celebrated the physicist David Bohm, who offered a cosmos that was internally related, totally interconnected, and self-regulating. Bohm referred to his vision of the universe as "wholistic" (Bohm 1981). What Lovelock and Margulis did for the Earth, people like Capra and Bohm did for the cosmos.

In the end Merchant saw the development of these open and dynamic areas of scientific analysis as positive, even if some expressions of them remained dependent on pre-revolutionary language and concepts. "Could Deep Ecology be cured of its antifeminist bias through greater sensitivity to its own language and analysis?" asked Merchant. "The answer is no.... The hegemony of capitalist over laborers depends on the exploitation of nature as a free gift to capital. The hegemony of men over women is necessary to maintain women's double 'second-shift' labor in the home and the workplace, whether in capitalist or state socialist societies. A science rooted in the twin assumptions of atomism and objectivity legitimated the domination of both nature and women" (Merchant 1992: 111).

Social and Political Philosophy

Merchant analyzed other movements, including spiritual ecology, social ecology, restoration ecology, Indigenous ecology, anti-globalization, sustainable development, and more, each of these exhibiting characteristics that were crucial ingredients for fomenting a post-industrial revolution with a socio-ecological order that was more just and sustainable. Some of these terms are used differently in different contexts. Her interpretation of spiritual ecology referred to building a partnership with nature that catalyzed a change in human consciousness (Merchant 1997: 18; *cf.* Sponsel 2012). Exemplars of this approach include Mary Evelyn Tucker and John Grim and the Harvard series on World Religions and Ecology (see the Chapter 4). Also included under the banner of spiritual ecology was the performance of a ceremony called the Council of All Beings. The ceremony was developed by Joanna Macy, a scholar of Buddhism who was an advocate for deep systems theory, in league with John Seed, an Australian rainforest activist. The centerpiece of the performance was a "talking circle" where humans came dressed as particular animals or species to speak from their perspective, to alert other animals to destructive environmental behaviors undertaken by humans. So, although we do not know what it is like to be a bat, or a cat, or a bear, we can imagine that the fragmentation of habitats detracts from bears' ability to maintain their normal range—their own fulfillment. Or take for example the rapid extension of disease vectors such as the white-nose syndrome, which endangered many bat populations. The Council of All Beings was widely used in environmental circles, particularly the more radical offshoots of mainstream environmentalism (Macy 2005; Taylor 2005).

Both Ruether and Merchant found capitalism to be one of primary culprits in the ecological crisis, and both traced the entwined histories of capitalist and socialist movements. Merchant briefly discussed social ecology, a field of study developed by the social theorist Murray Bookchin and his collaborator Janet Biehl (Clark 2005a, 2005b). Social ecology was imagined as an analytical tool for analyzing certain nested, power-laden relationships from the interpersonal, to the organizational, to the political. Bookchin was influential in libertarian circles, and more broadly was an advocate of anarchistic social arrangements. Anarchist philosophers were traditionally extremely wary of how religion and spirituality authenticated interpersonal and systemic domination. It is this thread that Merchant noted with apparent approval.

Bioregionalism is another approach she scrutinized; a political/economic model first called by that name at the 1972 North American Bioregional Conference. Political and economic arrangements were to be regulated only to the extent necessary so that most of the production and much of the economic activity took place at or below the level of the bioregion—an area based on biological and ecological zones rather than political and sometimes seemingly arbitrary lines that characterize modern-day states or provinces. Ideally, people *in* a particular bioregion do most of their producing and consuming within their own region, perhaps negotiating some measure of interregional exchange for necessary items (Snyder 1972; Sale 1985; House 1999; McGinnis 1999; Thayer 2003).

For instance, I live in North Carolina. Here we have a plethora of pigs, miles of chicken houses, textile production, mountainous forested regions for development and recreation, and on the coast, marine resources. North Carolina does not, however, grow avocados or citrus. If I want an orange, or an avocado I must look to Florida or California, but if I'm a bioregionalist, not to Peru, where most of the avocados in the grocery store were cultivated.[2] In a bioregional economy that simply means that I let avocados go—they are not a match for my region. An occasional orange might find its way here in a bioregional economy because of trade networks between bioregions, but the trade networks should have as little governmental interference as possible. Merchant discussed Deep Ecology and bioregionalism separately, and she believed bioregionalism still had something to contribute to the search for a sustainable human community. In practice, however, Deep Ecology and bioregionalism were closely related. Both were frequent topics in the pages of the *Earth First! Journal*. Often, they were discussed in conjunction, with Deep Ecology playing the role of moral philosophy, while bioregionalism acted as the political and economic philosophy. The scholar of religions Bron Taylor made a compelling case that bioregionalism was the de facto politics of the Deep Ecology movement, highlighting this pragmatic relationship between the two (Taylor 2000). The character of early radical environmental movements, led mostly by White men who unabashedly celebrated the manly virtues of living the "wild life," were received unfavorably by many ecofeminists, including Merchant.[3] All of these, however, were examples of radical ecologies (both conceptual and practical) that might mitigate the ecological crisis.

A Partnership Ethic

Merchant believed the revolution into a post-industrial sustainable world was best pursued through the application of an ethic of partnership between humans and the rest of nature. Merchant first offered her partnership ethic in her book *Earthcare: Women and the Environment* (Merchant 1996). The partnership ethic has been a staple of her work since then, typically offered in all of her 21st century works. She defined the ethic this way: "*A partnership ethic holds that the greatest good for the human and nonhuman communities is in their mutual living interdependence*" (*Merchant* 2003: 223,). She elaborated that "it is an ethic based on the idea that people are helpers, partners, and colleagues and that people and nature are equally important to each other" (223). It is worth asking what, if anything, this partnership ethic offers that is different from some eco-justice oriented and some organicist impulses. Merchant hoped that it would enhance collaboration, promoting a "discourse of cooperation" (224) to provide nonhuman nature with some additional operating space. Importantly for Merchant, the partnership ethic required a revolution in language, consciousness, and humans' obligations to nature. Some of the same, and some new scientific advancements were advertised as examples. Specifically, Merchant referred to postmodern science, by which she meant ecology, chaos theory, and complexity theory, which imagine nature as dynamic and always in process (*cf.* Marsh 1864, Leopold 1949; Plumwood 1995; Merchant 2003: 230). Merchant spelled out five precepts for sustainable partnership:

1 Equity between the human and nonhuman communities
2 Moral consideration for both humans and other species
3 Respect for both cultural diversity and biodiversity
4 Inclusion of women, minorities, and nonhuman nature in the code of ethical accountability
5 An ecologically sound management that is consistent with the continued health of both the human and the nonhuman communities (Merchant 2003: 224).

Partnership offers a compelling vision, but it is difficult to imagine what it looks like in action. In *Reinventing Eden: The Fate of Nature in Western Culture* (2013), as with her other works marked by the partnership ethic, Merchant provided some concrete examples of specific environmental contestations where a partnership ethic might be deployed to generate new and more sustainable solutions. Ultimately though, the partnership ethic "implies a remything of the Edenic Recovery Narrative or the writing of a new narrative altogether" (Merchant 2003: 242).

One potential critique that might be leveled is that, although Merchant celebrates diversity and uniqueness, the solution is a new grand narrative that glosses over differences to treat the categories of *nature*, *woman*, and *culture* in a way that perpetuates the outdated notion that they are separate to begin with. Is this partnership ethic derived from re-imagining Western science through the disciplines of *postmodern science* (ecology, chaos theory, and complexity theory) relevant only to women in Europe and North America? Or is partnership imagined as efficacious across the globe? Must women everywhere accept, or understand postmodern science before they can be liberated? These are difficult questions that Merchant did not entirely answer.

Merchant's book *The Anthropocene and the Humanities: From Climate Change to a New Science of Sustainability* (2020) addresses several humanities disciplines, shows how they have been impacted by climate change, and how they might positively contribute to a new sustainable world. Again, the sciences that she has in mind are the postmodern variety, ecology, and chaos and complexity theories, and what they tell us about our interdependence with the rest of the world, a world in constant flux. The remarkable breadth of her disciplinary travels includes treks through the disciplines of history (her home discipline), art, literature, religion, philosophy, ethics, and justice. The chapter on religion specifically offers a list of groups and organizations working to bend religious actors toward more eco-friendly habits (Merchant 2020). To her credit, in addition to the usual review of the so-called world religions, she noted Indigenous spiritual groups and interfaith groups that she claimed had positive impact. She celebrated Pope Francis's encyclical *Laudato Si*, "On the Care for Our Common Home," which advanced the idea that religion and science must band together to rectify longstanding injustices stemming from access to and distribution of resources. Merchant also mentioned the Earth Charter, which involved scholars of religions Mary Evelyn Tucker and John Grim. It was a blueprint for sustainable social arrangements complimented by the rescue of traditional ecological knowledge (TEK, see Chapter 5), a shift toward policy grounded in the Precautionary Principle, and a minimum common denominator social ethic. The inclusive and deliberative process of crafting the Earth Charter was an enormous undertaking that required consultation with governments, civil society actors, nonprofit organizations, and Indigenous rights and justice movements. Merchant emphasized that religions had an enormous role to play in combating climate change and publicizing the new holistic sciences. Despite the many statements from religious groups regarding better care for creation, there does not appear to be much evidence that people who identify with institutionalized religions are more environmental either in their attitude or their actions than the rest of the public (Taylor 2016; Taylor et al. 2016). Indeed, in some cases, religions may be obstacles to awareness of and affective relationships with nature. For some Christians, for example, this present earth is destined for destruction, which works against earth care in the present (Veldman 2014). In India some perceive the Ganges River as a goddess. But her holiness has invited environmental damage because many people believed the river could not be contaminated. At this writing it appears that the most environmentally ameliorative activities are undertaken by those who dwell on the peripheries of the world religions, or who are embracing new nature-based spiritualities (Taylor 2010; Johnston and Taylor & Johnston 2016; Taylor and Johnston 2016).

Glacken, Worster, and Merchant all reviewed the tension between organismic metaphors and tropes and increasingly dualistic and/or rigidified perceptions of nature over time. In all these tales there were subversive sciences which unlocked new descriptions of consciousness and associated shifts in the mode of production. Merchant specifically exposed the dominating hegemonic power structures, underscoring gender as a central variable, and pointed toward social movements of Indigenous and marginalized peoples, both in the global North and abroad. Whether she achieved her aims or not is a question left for the reader. But what is certain is that Merchant left a significant footprint in the field of religion and environment, particularly in drawing her analyses of scientific revolutions into conversation with the humanities.

QUESTIONS FOR DISCUSSION:

- What are the differences and similarities between the Colonial Ecological Revolution and the Capitalist Ecological Revolution?
- What is unique about Merchant's partnership ethic?
- How did Merchant relate the Scientific Revolution to environmental decline?

Notes

1 Environmentalist ire directed at the Glen Canyon Dam was popularized by the book *The Monkey Wrench Gang* (1975), which recounted the exploits of four fellow travelers who, overwhelmed by land development and governmental bureaucracies, turned to guerilla tactics to fight back, with the ultimate goal freeing the Colorado River from Glen Canyon Dam.

2 Ironically, I was informed by a sixth-generation peach farmer in Southern Georgia that citrus cultivation is already trickling over the state line from Florida to Georgia as climate change causes growing zones to migrate (personal communication 2019).

3 This rather masculine model was diluted in the 1990s when activists from Northern California and the Pacific Northwest joined in increasing numbers due to then President Clinton's Roadless Areas Review and Evaluation (RARE II) legislation (Tokar 1999).

7

WANGARI MAATHAI

"Becoming a Hummingbird" (Arboriculture)

The 2009 film *Dirt!* was a documentary about the living soil beneath our feet, featuring Wangari Maathai, the Nobel Prize winner, activist, and author who spearheaded the growth of the Kenyan Greenbelt Movement (GBM). In the film, Maathai recounts a folk tale about a hummingbird. There was a tremendous fire in the forest, and all the creatures were frightened, running for their lives. Even the mighty elephant and the proud lion fled from what they perceived must be certain death. Amidst the choking smoke, flames roaring at their backs, the animals cried out as they saw the hummingbird fly over. The tiny bird flew to the stream, gathered water in its beak, and dropped it resolutely onto the fire. Then he flew back to the stream and did it again. As he continued to fly back and forth, choking from the rising plumes, the other animals cried out to the hummingbird: "Hummingbird! What are you doing? You will be killed. And your beak is too small to carry much water, and the fire is so big—you cannot win!" And as the hummingbird flew overhead to once again refill its beak, he yelled back at the other animals: "I'm doing the best that I can!"

The lesson was about the role that even the smallest members of a community might play in the global drama of sustainability if given a chance. According to environmental activists the fire, of course, is an analog for the doomed trajectory of industrial civilization. The hummingbird might be interpreted as representing subsistence farmers and peasants, the people with little clout, of small size (economically and socially), but who possess the wisdom to adapt in specific places. In many cases, as Vandana Shiva, Rosemary Ruether, and others in this book demonstrated, oftentimes women were the keepers of horticultural knowledge and tradition. Today in many nations, ecological knowledge carried by Indigenous or traditional groups

DOI: 10.4324/9781003364801-8

is considered backward, or insufficient for more modern needs. Solutions from the global North, often technological or requiring technical familiarity, proved to be unsustainable, and they did not fit easily into existing worldviews (Harris 1966; Lansing 1991). Wangari Maathai was an important figure for those who study the intersection of religions and environments for at least three reasons:

1 The Kenyan GBM was largely led by women. They encountered obstacles from male civic and religious authorities. In many cases, even some activist's husbands perceived their wives' participation as a disruption of the natural order between God, men, and women
2 Her leadership is an example of twenty-first-century activism that stems from one of the "world religions"—Christianity—but must also contend with the lingering traditionalism still present in the surrounding culture
3 It is a good exemplar of a light green environmental religious movement.

Learning to Pull Together: Community Morality

Maathai's memoir painted a poignant picture of life before Christianity, arguing that "When European missionaries came to the central highlands at the end of the nineteenth century, they taught the local people that God did not dwell on Mount Kenya, but rather in heaven, a place above the clouds" (Maathai 2006: 5–6). Maathai was able to foreground forest protection in part because forests had at one time been important to the community. After all, trees count time in generations, not years. She was born when Kenya's transformation to capitalism was immature and incomplete. Men left rural areas for the cities to make money, staying away sometimes for long periods. As Maathai described, "It [capitalism] also introduced negative phenomena such as prostitution, absent fathers, and sexually transmitted infections that were unknown until then and persist as significant challenges in society today" (Maathai 2006: 14). Maathai was born into the colonized terrain of economic globalization, but still valorized the old ways as more connected with place. Meanwhile, she lamented that in her own hometown,

> these plastic [bags] litter the parks and streets, blow into the trees and bushes, kill domestic animals (when they swallow them inadvertently), and provide breeding grounds for mosquitos. They leave the town so dirty it is almost impossible to find a place to sit and rest away from their plastic bags.
>
> *(Maathai 2006: 35)*

Global markets met demand by liquidating resources for development and profit. Colonialist arborists said that forests could provide more income if agricultural and other land was transformed into cash crops of trees. Nonnative species like pine trees and eucalyptus could turn a profit, but "they eliminated local plants and animals, destroying the natural ecosystem that helped gather and retain rainwater" (Maathai 2006: 39). The quick-growing nonnative species spewed forth a canopy that shaded out traditional trees. Maathai emphasized that her parents and grandparents were naturalists, "without conscious or deliberate effort, [their] cultural and spiritual practices contributed to the conservation of biodiversity" (Maathai 2006: 46). The colonial impacts in central Africa were overtaking those traditional perceptions.

During the administration of U.S. President John Kennedy, Maathai's scholastic talents earned her a scholarship when Kennedy urged the State Department to open to more African students (Maathai 2006: 74). Kennedy himself provided the financing for the award through his family foundation. Maathai completed her degree in 1971, recounting proudly, "I was the first woman in East and Central Africa to receive a doctoral degree" (Maathai 2006: 113). Particularly in Africa during that time it was a remarkable achievement.

In 1962, roughly a decade before Maathai started her activist career, the marine biologist Rachel Carson wrote her famous book *Silent Spring*, blowing the whistle on chemical manufacturers and their products. The chemical corporations in turn launched a massive public relations campaign to discredit Carson's research, but also impugn her character. In Maathai's case, she blew the whistle on corrupt government officials who formed alliances with Western extractive industries. She was subjected to the same sort of character assassination and misogyny. Maathai's work with women and on sustainable livelihoods preceded her participation in the GBM, which was her most well-publicized success.

Her early activism included the formation of other woman-led environmental groups, including the group *Save the Land Harambee*, an organization dedicated to implementing sustainable development in the African context.[1] Working with women in particular villages she catalogued the survival challenges that they were facing. For most of the women she engaged, there were some commonalities. First, wood was required to cook any food they grew. But there was no fuel wood. Second, there was increasing desertification across the region where Maathai was working. Women reported having to walk further to obtain water resources. Third, as middle- and upper-class people in Kenya began to emulate Western diets, including processed foods, they had less use for locally raised vegetables and livestock. The poorest who grew the foods that for generations had been subsistence for whole communities, could not sell their crops. There they were, with food they could not cook because of a lack of fuel wood; food they could not sell because consumer preferences leaned toward processed Western food; and with what meager funds they could scrape together they bought the scraps of those processed foods from across oceans to survive (Maathai 2006).

In 1976 Maathai traveled to the U.N. conference on Human Settlements (HABITAT 1) in Vancouver. She was "bombarded" with ideas while there including "the need to appreciate that planet earth is our only 'ship'" and that we therefore needed to chart a morally sound course. She called for "a halt in the over-exploitation of resources, the need to control the development of cities to prevent the creation of 'concrete jungles,' and the problem of air pollution through road transport" (Maathai 2006: 15). Interestingly Maathai referenced the "spaceship" metaphor which was commonly deployed in environmental literature. It can be traced to Kenneth Boulding, an environmental economist, who referred to "spaceship earth" as a metaphor for the politics of navigating a growing population with a finite pool of resources. Garrett Hardin, an economist who was often called a "neo-Malthusian," critiqued and extended Boulding's metaphor. Hardin claimed that Boulding's spaceship was not an appropriate metaphor for dwindling resources, since a spaceship (or any ship for that matter) has a captain, not just a gaggle of people who chart a course by committee.[2] Likewise, Maathai argued that the world ship required "ethical steering" (Maathai 2006: 15).

But her vision was distinct from Hardin's. Hardin advanced the now-famous idea of the tragedy of the commons (Hardin 1968). According to Hardin, if, say, three shepherds grazed their livestock on a common pasture, each one would be motivated to maximize their own use of the resource. The *benefits* of *over*grazing accrued to each individual shepherd, while the *cost* of maintaining the commons was borne by all. The result was the inevitable destruction of the commons because of a race to maximize individual utility from common goods. The only solution to the tragedy of the commons, Hardin believed, was privatization of the commons, which would preclude any such contestations in commons regimes. Hardin encouraged "lifeboat thinking" rather than a spaceship metaphor (Hardin 1974). The captain of each lifeboat (a metaphor for nations) had an existing stock of resources which must be used sustainably by managing the lifeboat population. If there were too many people on the lifeboat, it would surely sink. Hardin's metaphor imagined the industrialized countries (the global North) as the lifeboats, while the global South were merely swimmers in the water, dependent on any scrap of mercy from those who commanded resources.

Hardin made a compelling if controversial case for speaking openly and honestly about population as a driver of the extinction crisis. However, there were questionable fundamental presuppositions that guided his theory. First, his thought experiments with the commons and lifeboats depended on a specific anthropology—a vision of how a person exists within an economy as a function of their own preferences, and why. Hardin assumed that the unit of analysis was the individual and that all individuals had a utilitarian and transactional worldview. He imagined that the economy was comprised of bounded, discreet, cost-benefit calculators who acted only for their own individual good. Second, Hardin assumed that each of these individuals inherently wished to maximize their own individual good. But, these assumptions were not universal features of human decision-making. People-in-community typically do not make decisions without regard for those who share their place and lifeways (Ostrum 1990, 1994). Moreover, people do not always act rationally. That is, they do not blindly pursue only their own benefits, *especially* when sacred values are at stake (Atran et al. 2002; Atran and Medin 2008; Atran et al. 2005Atran & Medin 2008). Counter to some of Hardin's assumptions above, Maathai's community-building work celebrated the cultivation of an extended sense of responsibility and group efficacy. This was, in fact, one of her primary contributions to the environmentalist milieu.

Maathai worked for more than 20 years in this capacity in her activist life as the leader of the National Council of Women of Kenya, supported by European organizations, and eventually funds from the United Nations Development Program (UNDP). She was present at the United Nations Conference on Environment and Development (UNCED) and the "Earth Summit" in Rio de Janeiro. All the while she was also working to support the National Council of Women of Kenya, cultivating the seeds of the increasingly important Greenbelt networks she developed. She was also one of many involved in the creation of the Earth Charter, and an advocate for its aims.

The Affective Habitat of the Kenyan Greenbelt Movement

First as part of her NCWK work, and later separate from it, Maathai developed the Kenyan GBM. What she had started on a very small scale in 1977 grew through her hard work and eventually, into her international persona. The campaign was grounded in the awareness of the struggle for survival encountered mainly by women in Kenya. Phase I of the Greenbelt campaign aimed to facilitate people's understanding of the environment as something to be valued, to encourage a felt moral obligation toward the Earth. The GBM stated that they were "a values-driven society of people who consciously work for continued improvement of their livelihoods and a greener, cleaner world."[3]

The GBM was a values-driven movement, and one of its essential goals was to export those values into other African contexts. Their mission was "community consciousness for self-determination, equity, justice, environmental conservation and improved livelihood securities (food, shelter, education, health, employment, and human rights), using both civic and environmental education and tree-planting projects as the entry point" (Mburu 2005).

Specifically, all the members of the group worked to embody the following five values:

1 Love for environmental conservation
2 Self and community empowerment
3 Volunteerism
4 Strong sense of belonging to a community of Greens
5 Accountability, transparency, and honesty (Maathai 2006: 33).

A more recent rendering appears on the GBM's webpage where the core values are the same except for the removal of the fourth—a sense of belonging to a group of Greens. While there is no public explanation for the removal of the fourth value, it is possible that it was associated with a specific political party, while the larger group wanted to be broadly inclusive.

Activists carried these character traits into their advocacy, which combined civic education and tree planting campaigns to increase food security at the household level (Maathai 2006: 34). The tree plantings provided a sustainable fuelwood source for locals. Through her connections, Maathai funded the purchase of native seedlings which were distributed through local networks, facilitated by agents from the GMB. These agents were required to check on the status and health of each tree one month from planting and three months from planting, to gauge the success of the strategy. The agents of the GBM were paid a modest wage for this extension work, and they enrolled in related classes and workshops on organic farming, crop rotation, Indigenous plant ecosystems, and farming techniques (Maathai 2006: 33–54). It was an almost magical formula: planting the trees gave women not only access to fuel wood, but also decreased the distance women had to walk to get it, *and* the tree roots helped prevent erosion. Some workshops included the role of government, which according to Maathai was largely either inept or corrupt. But in either case, the technical knowledge and expertise of the government was not shared with the local populations.

While there was indeed high-level governmental support for Maathai's initiative, it eventually ran aground because of two developing obstacles: (a) the male-dominated political party saw in Maathai's assertive leadership a threat, whether implicit or explicit, against the patriarchal, top-down style of governance; and (b) any on the ground communication from the government was couched in useless jargon and methods that did not translate to the largely female cohort who were trying to enact sustainable lifeways at the household level. As Maathai put it, "Unfortunately the foresters insisted on using technical terms to describe the specialized tools and inputs needed to run a successful tree nursery" (Maathai 2006: 27). One might also question whether those foresters had much investment in teaching poor women how to create a sustainable form of agriculture, given the androcentric milieu. In any case, this prompted a "revolution": "The women decided to do away with the professional approach to forestry and instead use their common sense! After all, they had for a long time successfully cultivated various crops on their farms" (Maathai 2006: 27). Maathai elaborated, "The campaign encouraged them to use their traditional skills, wisdom and plain common—and perhaps women—sense" (Maathai 2006: 27–28). Maathai recalled proudly that she had developed an army of "foresters without degrees"—utilizing traditional knowledge instead of technologized management. The values-driven movement depended on an experiential, and deeply felt connection with place, and concomitant obligations to those places.

The Fruits of the Green Belt Movement

The GBM was comprised of a grassroots network of peasant farmers—both women and men. The movement de-centered the role of government in protecting the environment and the livelihood of people, instead centering the traditional ecological knowledge of villagers, especially women. The balance of the population in Kenya remained rural.[4] At the turn of the new millennium, the GBM claimed to have established 6,000 women-led tree nurseries throughout Kenya. The women directly affected by the movement reached 100,000 female farmers. The intent was for those women to be the spark that ignited the rapid spread of life-affirming values and locally sustainable practices. That was the essence of Phase I of their strategic goals (1977–1999). Once the volunteer network was established, Phase II included the distribution of compensation paid to women farmers who planted on public lands (Mburu 2005).

The GBM model and its export to other countries in Africa resulted in a sort of Pan-African Green Network. By 2005, the GBM was established in 15 countries on the continent. After the compensatory process, Phase II shifted toward the preservation of biodiversity and ecosystems, something broader and more inclusive than tree planting. The low-hanging fruits were easy to see—women did not walk as far for fuel wood, and village nutrition improved because of the presence of fruit trees. But it was more difficult to see that beneath the surface were rich root structures that secured the soil and prevented erosion. The roots and soil acted as a sponge, and increasing plant fertility mitigated desertification. Women who were responsible for food cultivation also became local experts in Indigenous crop and seed preservation, locally appropriate crop rotations, and the grassroots development of social cohesion. It was a sense of belongingness, of group identity, that was not dependent upon any political party or agenda. It was unified by a worldview which included life-affirming values and the responsibility to act on them. Maathai listed the GBM's achievements from Phase I:

1 Rural forestation and reforestation
2 Creation of employment within the movement
3 Raising awareness of the importance of the environment
4 Empowerment of individuals and communities
5 Community mobilization
6 A source of inspiration
7 Raising awareness of the need to protect local biological diversity of plants and food crops
8 An improved image of women
9 Increased advocacy and networking activities
10 Increased number of agricultural tools in communities
11 Extensive documentation and recognition of the Green Belt Movement
12 Survival of the Movement (Maathai 2006: 63–64).

Each of these was significant in the context of poor rural farmers. Taken together, they are a revolution. Interestingly for our purposes here, the GBM provides a case study of the cultivation of green spiritualities. One of the goals of the movement was to imbue individuals with the idea that the Earth has some inherent worth, or intrinsic value. Other goals included very pragmatic improvements in subsistence farming techniques.

There was a dynamic in Maathai's writing and activism that betrayed some ambivalence about her Christian identity. On the one hand, she was from a culture that was 85% Christian, and that religious identification was hers since birth. She was partially educated in religious educational institutions, and her activism was publicly deployed in Christian tropes and metaphors. Yet she recognized colonialist Christianity as the main force which eradicated the traditional African sense of belonging to land, and to other-than-human creatures. These commitments to land, and to non-human animals, were the part of the traditional African worldview that Maathai wished to preserve. There was a tension there. In some contexts, traditional wisdom was actively suppressed as a holdover from a supposed pre-Christian depravity. Maathai's ambivalence was understandable given the pull from both directions. Maathai's experience was one shared across the nations where the GBM took hold, where there were many women in such circumstances, beholden to both the confines of Christianity *and also* to traditional lifeways and sources of inspiration. As the scholar of religions, Graham Harvey once noted about a neighboring African country, Nigeria's official demographics revealed the population was 50% Christian, 40% Muslim, and 10% traditionalist. But if you looked closely, behind the icons and significations on the exterior of a building, on the inside one might still find shrines for local preternatural beings or amulets for traditional healing. It may lead you to the conclusion that the real demographic breakdown is better visualized as 50% Christian, 40% Muslim, and 90% traditionalist (Harvey 2013: 113).

The Specter of Religion and the (Light) Green Spirituality of the GBM

Recall that in E.B. Tylor's purview, religion was prompted by the perception of a specter, or soul, outside one's body. But, the specter appeared in the profane world, also; the everyday negotiation of living depended on the conciliation of long-dwelling spirits of place. In essence, the vertical relationship between humans and a deity was satisfied by Christianity. But the horizontal relationships—those between humans or between humans and other carbon-based lifeforms—were important for procuring subsistence, and for the mediation of delicate social exchanges. Oftentimes when situations were handled according to the old ways, they might be dressed in a Christian disguise.

This illuminated a conundrum: most broad survey instruments have assumed that humans are "boxable," that their identity and the performance of it are mostly related to this or that identity marker, whether it is religious, gendered, ethnic, etc. Unfortunately for some surveyors, such categories compromised the precision and the generalizability of their findings. Fortunately, for us it means that the world is comprised of much more interesting and interactive agents. Most people are religious, intellectual, and practical hybrids who are just temporary amalgamations of multi-layered, multiply identified, and cognitively fluid persons. Instead of a classificatory dupe where religions (and moral sensibilities, and ethnic divisions, etc.) are mutually exclusive, perhaps we ought to find more varied and stimulating networks of beings, only some of whom were human. But what is to be gained by making the world *more* complex instead of less? Must we jettison all classification and categorization as wrongheaded? Of course not. Our own unique biases as cultural detectives are revealed as we engage in negotiations with cultural, ethnic, and ethical "others." Those forced choice classifications are doing real cultural work not because they are universal, but rather because they are ethnocentric concepts portrayed as *natural*—something that just goes without saying. The question, then, is not whether binary classifications are useful. Rather the question is for whom are they are useful, and subsequently in what contexts (cultural, political, religious) are they exercised?

When people identify as *religious*, they often imagine an institutional commitment and function. In contrast, when people refer to *spirituality*, they refer to more informal, eclectic, combinatory personal perceptions and practices. Many social formations referred to as spirituality evidence nature spiritualities, ecopsychologies, social ecologies, and much more. It is not useful in the context of this book to clarify definitions. Making distinctions like these may be important in other scholarly contexts, but in general the difference between religion and spirituality is characterized this way (Johnston 2016, 2018; Johnston & Taylor 2016).

Such distinctions are important, but in no case is the classification "religious" or "spiritual" something that can be universally applied. Classifications are, as J.Z. Smith noted, products of scholars' own interests and relevant to the sorts of questions they hope to investigate (Smith 1998). In the case of the Green Belt Movement, the performative part of activism was Christian, while its foundations, its values, were Kenyan, or even Pan-African, as Maathai's influence beyond Kenya illustrated. African Christianities asserted human superiority over other creatures and assumed that the use of resources was a human prerogative. The GBM struggled with reconciling the dualistic metaphysics of Christianity with long-standing, place-based propitiation of spirits. The GBM and similar movements should not be characterized as dark green religions. The scholar of religions Bron Taylor analyzed the GBM and found that many if not most of its advocates neither fully embraced a Darwinian understanding of evolution nor an appreciation of the importance of biodiversity, and thus concurred that they were not dark green religions (Taylor 2013). Even if GBM's behaviors resulted in improvements in biodiversity and preservation of ecosystem services, according to the dark green ethos, it did not represent a full-blown dark green social movement. Rather the GBMs are light green environmental religions which ground their species-centrism in something more like stewardship than dominion.

Is a thorough (or even surface level) understanding of the full implications of Darwinian evolution necessary to perpetuate the GBM, or other localized responses to climate catastrophe? This is of course debatable, and the pragmatist might observe that in either case the outcome—the protection of biodiversity—is positive. But it is also true that a fuller acceptance of the implications of Western science would illuminate, for instance, connections between population and environmental degradation, something that is notably absent in most African Christian Churches. In any case, it is probable going forward that exclusivist religions which do not or cannot endorse the global reach of Western science will go extinct over the very long term. Not because Western science is "more correct," but because "science" was deeply enmeshed with neo-colonialist sustainable development schemes and a phase of corporate globalization which overtook much of the world in the twentieth century.

In 2004 Wangari Maathai was awarded the Nobel Peace Prize for empowering women and enhancing the resilience of communities in East Africa. Specifically, she drew explicit connections between democracy, environmental protection, and gender inequality in Africa and beyond. The GBM was a religious environmental movement: it was carried out under the auspices of a Christian ministry, but beneath the surface it belied a practical integration of traditional farming methods with new technologies. Christianity, like all the Abrahamic religions, historically exhibited an anthropocentric bias. But it was a mode of Christianity informed by local traditions, which often did not have a word for *nature* as separate from humans. In the *Encyclopedia of Religion and Nature*, an entry by Mburu (2005) was accompanied by a passage from geneticist and environmentalist David Suzuki:

> "All of us have a God in us, and that God is the spirit that unites all life, everything that is on this planet. It must be this voice that is telling me to do something, and I am sure it's the same voice that is speaking to everybody on this planet...*Poverty and need have a very close relationship with a degraded environment ... to break the cycle, one has to start with a positive step, and I thought that planting a tree is very simple, very easy – something positive that anybody can do*" (Suzuki 2022 [1991]: 222, 223).[5]

Planting a tree is easy. But the patriarchal, exclusivist monotheistic society has its own inertia, and changing its trajectory is tedious and time consuming.

To sum up, all the classifications at play in this chapter—Christian and traditionalist, government overtures versus local action, and a women's movement—had material impacts. Even as there were contestable terrains related to African Christianity and traditional African religions, they could work together to generate conservation behaviors. The life of Wangari Maathai dramatically changed the landscape of African nations. Her philosophy of women-inspired grassroots organizing provided increased social accountability, focused the desire of peasant constituencies, and channeled economic exchange in important ways, and in that she was a hummingbird, doing environmental religious work.[6]

QUESTIONS FOR DISCUSSION:

- How would you explain in your own words the tensions between different religions exposed in Maathai's story?
- How did Maathai's embrace of an "Earth democracy," to use Shiva's term (see Chapter 9), challenge existing power structures and historical gender narratives?
- Why would Maathai's initiatives qualify as light green environmental religions?

Notes

1 The term *harambee* means *let us all pull together* (Maathai 2006: 20f).
2 Thomas Malthus (1766–1834) was an economist who first argued that human population growth would eventually overcome the Earth's capacity to

support it (an early articulation of the concept of carrying capacity) (Robin et al. 2013: 15–62). But Malthus's controversial approach suggested as a solution the sterilization of destitute populations. Hardin did not support this specific solution, but his work was focused on population growth, what he called the "wild taboo" (Hardin 1973). Historically several religious groups and institutions have resisted any serious discussion of the contribution of population to environmental degradation or other social issues

3 From the GBM (http://www.greenbeltmovement.org/who-we-are), accessed 19 June 2023.

4 CIA Factbook (https://www.cia.gov/the-world-factbook/countries/kenya/#environment), accessed 19 June 2023.

5 The italicized portion of the quote was attributed to Priscilla Sears in Suzuki (2022 [1975]). The quote in the *Encyclopedia* was actually cobbled from two different pages in Suzuki's book (2022 [1975]: 222, 223). Reviewing Sears's piece, titled "Wangari Maathai: 'You Strike the Woman…,'" it is clear that only some of the quoted material appears in the actual document. It is unclear whether the rest of the words were erroneously drawn from a source other than Sears (1989), or whether they were a creation of Suzuki's.

6 I am drawing here on the scholar of religions David Chidester, who noted that something is doing religious work when it draws communities of people together, focuses desire, and facilitates exchange relationships in particular directions (Chidester 2005: 6).

8

J. BAIRD CALLICOTT

"From the Green Fire to Worldview Remediation" (Environmental Philosophy)

Imagine loosening your belt after a holiday meal as you stroll over to the nearby prairie to help light a prescribed burn. This is the sort of "ritual of restoration" that ecologist William Jordan III suggested the ecological hero Aldo Leopold would value.[1] J. Baird Callicott (b. 1941) was the most well-known exponent of Leopold and his conservation vision. He received his PhD in Philosophy in 1971, and the same year taught the world's first environmental ethics course at the University of Wisconsin-Stevens Point. Probably his most well-known contribution was to provide a systematic elaboration of the ecologist Aldo Leopold's Land Ethic. One of Leopold's famous dictums in his posthumously published *A Sand County Almanac* (1949) was: "a thing is right when it tends to preserve the integrity, stability, and beauty of the biotic community. It is wrong when it tends otherwise" (Leopold 1949: 262). Callicott interpreted Leopold as advancing an ecocentric ethic—one in which the land itself and the ecosystems contained within it were the unit of moral considerability. To do this, he had to contend with the emergent animal rights and animal liberation advocacy. The first fundamental component of Callicott's legacy was the argument for a non-anthropocentric ethic—one not dependent upon human valuation to legitimate the worth of nature. It was an easier but insufficient task to imagine an anthropocentric—human-centered—ethic.

How the Land Ethic Informed Twentieth and Twenty-first Century Environmental Ethics

Aldo Leopold was among the first ecologists in the United States after Yale University established the first School of Forestry under the leadership of

DOI: 10.4324/9781003364801-9

Gifford Pinchot, who had been educated in European schools. Leopold was, like most of his cohorts, pragmatic when it came to land use and resource exploitation. But Leopold shared in his posthumously published *A Sand County Almanac* (1949) a story that shifted his whole perspective—an emotional eye-to-eye connection with a non-human animal. He had shot but not killed a she-wolf who was with pups and arrived beside her broken body in time to "watch a fierce green fire die" in her eyes. It was this instance where Leopold is considered by many to have offered the first truly "ecocentric" ethical perspective. While hunters wanted to exterminate wolves to preserve both domesticated animals and game, that perspective was not shared by the wolf, or by the mountain upon which the incident occurred. It was this broadening of the scope of his moral affiliations that led him to the notion that landscapes, even whole ecosystems might have their own "good."

Importantly, the development of his seminal Land Ethic depended on the rejection of Abrahamic religious traditions and the development of a new morality, which perceived humans as a part of the ecological system, and the system as the unit for ethical analysis. Leopold's writings and life's work, and especially his story about the green fire, became an organizing motif for the radical environmentalist milieu, and of conservation biologists. Curt Meine's brilliant biography of Leopold (Meine 2010), and a later film called *Green Fire* (2019), have become familiar for those who study the intersections of religions and nature. In fact, what many consider the first dissertation written about the emergent field of environmental ethics was J. Baird Callicott's elaboration of the Land Ethic and its application in the protection of the legacy of wilderness in the United States.

The field of environmental ethics emerged as scholars began ruminating on questions such as: (1) should ethical deliberation about the natural world only be centered on humans, or should it include other-than-human entities? (2) If it should include non-human creatures, which ones count in our moral reckoning? If ethical deliberation goes beyond the human realm, should humans concern themselves only with organisms that are beneficial to our own well-being, or should we care for and about those with no direct utility to us? (3) A related question is if non-human persons are relevant to our moral reasoning, do they count the same? (4) Can we envision one coherent ethical system that encompasses all creatures of concern, or is a situational ethical framework best?

Both scientific perspectives and organismic arguments for respecting the natural world continue to influence the public imagination. In 1970, the first Earth Day took place. In 1971, the first conference dedicated specifically to environmental ethics was organized by Frederick Ferre and Eugene Hargrove at the University of Georgia. Those two would later launch the journal *Environmental Ethics*, with Hargrove later taking it with him as he moved to the University of North Texas. Articles and books dedicated to the topic became more frequent, and one of the most pressing questions, it seemed, was the unit of moral considerability

Callicott's approach was a monist, non-anthropocentric—specifically eco-centric—ethic grounded in the intrinsic values of the ecosystem as advanced by his inspiration, Aldo Leopold (1949). To say that his ethical perspective was eco-centric means that it was not just carbon-based creatures that were of concern. Instead, functioning ecosystems were the unit of moral considerability. The notion of intrinsic value referred to entities having ethical worth aside from any human utility or valuation. Aldo Leopold, his hero, endorsed the idea that the American conservation movement was in trouble because it still operated according to old religious traditions that promoted human exceptionalism. Callicott concurred. In the early to mid-twentieth century, most conceived of humans as higher order beings, above other creatures, and certainly more important than plants. What was missing was a sense of respect for the land. The prevailing perception of land was both cognitively and biophysically separated from the human realm. Thus, there was no impetus to care for the land. "There are two spiritual dangers," Leopold warned "in not owning a farm. One is the danger of supposing that breakfast comes from the grocery, and the other that heat comes from the furnace" (Leopold 1966 [1949]: 6). Humans too often did not recognize their dependence upon other animals, and on the primary producers who, through photosynthetic alchemy turned sunlight into usable energy. The dualism between humans and environments was constructed by centuries of (mostly) religious actors who had their reasons for discounting the value of land. As Leopold put it, "Conservation is getting nowhere because it is incompatible with our Abrahamic concept of land. We abuse land because we regard it as a commodity belonging to us. When we see land as a community to which we belong, we may begin to use it with love and respect" (Leopold 1966 [1949]: xxii). Callicott articulated this vision more comprehensively than anyone and wrestled with moral decision-making through that ecocentric lens.

For Leopold, the gradually expanding idea of "personhood" was articulated in what he referred to as the ethical sequence. Step one was a moral code that dictated relationships between individuals. The second step involved the relationship between the individual and the society. The third, not yet taken, was to form an ethical relationship with all of the buzzing, breathing, clicking, clacking, stinging, chirping, and howling elements of the ecosystem.[2] The relationships between them all was what Leopold referred to as the "land." As Leopold highlighted, "there is as yet no ethic dealing with man's relation to land and to the animals and plants which grow upon it" (Leopold 1989 [1949]: 190). Aside from Leopold, the emergence of this ethical perspective was nascent in the conservation movement, but not yet realized. But as Leopold noted, "No important change in human conduct is ever accomplished without an internal change in our intellectual emphases, our loyalties, our affections, and our convictions. The proof that conservation has not yet touched these foundations of conduct lies in the fact that philosophy, ethics, and religion have not yet heard of it" (in Flader & Callicott 1991: 338; also quoted in Meine 2005). The development of an *ecological conscience* was the goal. Posthumously, Leopold became a hero for the environmental movement and one of the main protagonists (if not *the* central figure) in Callicott's significant written corpus.

Callicott religionized science. He used concepts and findings at the leading edges of scientific inquiry, and specifically articulated how such findings might inform humans' fundamental moral stance. The historian of science and technology Lynn White, Jr. (1967) wrote an influential article in *Science*, one of the most widely read journals in the Anglophone world, which put the bulk of the blame for the ecological crisis on the backs of what he referred to as the Judeo-Christian traditions. Specifically, human dominion was the lynchpin of all Western ethics and religions. A theological justification, allied with increasingly invasive agricultural technologies, produced startling ecological impacts. Whether White was right or wrong, the social function of human exceptionalism was at least in part to destroy pagan social formations which identified the holy with the Earth or its denizens. If the cause of the ecological crisis was so obviously religions, thought White, then so must the solution be. Leopold clearly presaged White's later impactful work, which sent scurrying several theologians determined to mine their traditions for possible alternative translations regarding humans' place in the rest of creation. Mining green glimmers in ancient texts and attempting to find environmental wisdom within existing religions became an active area of scholarly production (Santmire 1970, 1985; Schaefer 1970; Swimme & Berry 1992; Tucker & Williams 1997; Tucker & Berthrong 1998; Berry 2000; Chapple 2000, 2002; Girardot, Miller & Liu 2001; Grim & Tucker 2014; Johnston 2013, 2014, Taylor & Johnston 2016). Francis Schaeffer's *Pollution and the Death of Man* (1970) and other similar works advanced an ethical stance informed by theology rather than philosophy. Schaeffer's work (and later theological approaches, for instance, Wilkinson (1980) and Gustafson (1996)) argued that White's thesis appeared to have some merit. However, according to these figures, Christian mistreatment of the environment was really a result of the *misinterpretation* of the biblical message. Ethicists such as John Gustafson (1996) referred to their approach as theocentric rather than anthropocentric. Contrary to bio- or eco-centric positions, theocentrists argued that what is good for humans ultimately derives from obedience to divine commandments, not from rational human approaches to expanding ethical consideration.The starting point for ethical deliberation ultimately derived from acceptance of specific notions of God. Theocentrism is, in theory, a non-anthropocentric ethical stance. If the propitiation of an ostensibly male deity who sits atop the great chain of being, with humans above all the other creatures and little less than God (and in this case, men were also "higher" than women), then whatever theocentric reasoning occurs will cash out into anthropocentric practices and rituals.

If our goal as cultural critics is to develop the ability to analyze and critique culture, we may do well to adopt a functionalist perspective, which brackets off theological explanations and focuses on the very real (and empirical) economic, political, and social impacts of such claims. A functionalist explanation focuses on observable political, social, and economic impacts, and is valid whether or not nonempirical theological positions are true. Whether overtly human-centered philosophical approaches or theocentric approaches that justify human exceptionalism with nonempirical entities (gods), both approaches privilege humans at the expense of the rest of the ecosystem. The non-human world was a repository to be unlocked when humans desired it to be, to exploit or liquidate resources for economic gain.

Non-anthropocentrism and Intrinsic Value

One easy point of access to questions regarding human-centered ethics is William Baxter's influential work *People or Penguins: The Case for Optimal Pollution* (1974). Baxter argued for a thoroughly human-centered (anthropocentric) ethic wherein there was a level of pollution that was optimal for human flourishing, and that human satisfaction should trump the "good" of other-than-human species. If human-generated pollution compromised the ability of other species to survive, so be it! The value of other-than-human species—to the extent that they had any—was purely aesthetic and relative to human enjoyment. As Baxter put it:

> My [ethical] criteria are oriented to people, not penguins. Damage to penguins, or sugar pines, or geological marvels is, without more, simply irrelevant. One must go further, by my criteria, and say: Penguins are important because people enjoy seeing them walk about on rocks.... In short, my observations about environmental problems [are] people-oriented, as are my criteria. I have no interest in preserving penguins for their own sake.
>
> *(Baxter 1974: 5)*

The first criterion for Baxter's ethical rumination is what he referred to as "spheres of freedom": "every person should be free to do whatever he wishes in contexts where his actions do not interfere with the interests of other human beings" (Baxter 1974: 2). Second, waste is "a bad thing" (Baxter 1974: 3). Third, all humans should be regarded as an end in themselves, rather than as a means for the betterment or enjoyment of another human (Baxter 1974: 4). And fourth, every individual should enjoy the opportunity and incentive to improve their share of satisfactions (Baxter 1974: 4). So, here was a purely human-centered argument advancing an almost libertarian perspective on the pursuit of human well-being. By this rationale, provided it did not impinge on another *human's* dignity or well-being, I would be welcome to chop down the last tree in a forest, to mine the last morsel of coal from a mountain (after decapitating it), to extract the last bit of oil from a fragile ecosystem by drilling, or even (to take the examples to their logical conclusion) shoot and kill the last of a species to hang as a trophy on my wall, so long as it brought me joy, and did not impinge on any other humans' sphere of freedom.

Of course, such an ethical sequence raises some important questions. Imagine shooting the last remaining White Rhino to enjoy the hunt and the trophy, or to sell its horn for medicinal purposes, because it increased one's own economic happiness. Would such an act not diminish the ability of other people to enjoy seeing rhinos in the wild? Or to use a more pertinent example, would killing off all the penguins compromise the sphere of freedom of others who *do* like to see them walk about on rocks?

There were scholars already pursuing the rights of non-human animals and their liberation from the confines of industrialized food production. Baxter was responding to some of these flowering animal arguments with his human-centered stance. Some of the most noteworthy animal rights advocates, however, advanced a *biocentric individualism*. That is, the unit of moral considerability was imagined as extending to non-human living things, typically grounded in the articulation of some level of sentience or self-determination. One of the most important articulations related to animal ethics was the philosopher Peter Singer's work *Animal Liberation* (1973), in which he advanced an ethic that extended moral considerability to all sentient beings. Of course, the question is, who counts as "sentient"? Singer's impact was significant, prompting a flurry of responses. Singer's was a *utilitarian* argument for the protection of animals—one which advocated the greatest good for the greatest number of ethically considerable beings.[3] Singer drew the boundary of moral consideration and sentience somewhere around the phylum *mollusca*. The suffering of individual animals was of primary importance in this model.

Tom Reagan's *The Case for Animal Rights* (1983) answered with a biocentric individualist approach with a *deontological* bent. In other words, there were rules for how humans were to engage with what he referred to as "subjects of a life," those who were competent enough to pursue their own ends, their vision of what was "good" for them. In the end, Reagan's work really ended up being something closer to a "mammal rights" than an "animal rights" argument, since the animals who were capable of such self-centered deliberation are those that are, or are most like, mammals. Both Singer and Reagan worked toward a *biocentric* ethic, one which included other-than-human animals, and in both cases the primary concern was with individual creatures (not species or landscapes).

Callicott (1980) criticized both animal rights and animal liberation arguments. The former (typified by Reagan's approach), he said, stemmed from *ethical humanism*, the limitation of moral considerability to those animals that share certain higher functions and characteristics with humans. Animal liberation (Singer's approach), he suggested, stemmed from a preoccupation with *humane moralism*, also an approach that was concerned, ultimately, with the suffering of *individuals*.

When I lived in the U.S. state of Wyoming, I was the victim of a long-term siege from a mischief of magpies who dwelt near my house.* They overturned trash cans, ripped the bags open to find morsels, and scattered any inedibles across the yard. I tried several deterrents (I assure you no magpies were injured in the real-life occurrences related to this story). None of the measures I introduced kept the magpies away. The magpies and I, we had a conflict of interest. We were at an impasse. Should I destroy the birds? I knew they were just doing what magpies do to survive. And to eliminate them from the habitat seemed, to me, like a step too far. On the other hand, conservation biologist Michael Soulè noted that his own Buddhist practice obligated him to shoot the feral cats that roamed around Paonia, CO, his home. Those cats were not native to the habitat. The birds were.[4] In both cases, the actors decided based on ethical reasoning. Interrogating whether those decisions were *right* or *wrong* depends in very large part on the situation and the collective norms of their respective communities of accountability. I thought the lives of the birds were worth my restraint, although I held on to some resentment. I exercised a sort of biocentric individualism. Soulè, on the other hand, was seized by an ethical obligation to assassinate cats to restore balance to the habitat, working from an ecocentric sensibility. The differences matter.

Do such approaches raise more questions than they answer? By the standards of biocentric individualism does the moral evaluation include only creatures who think like us (humans), or those we deem "rational," or those we think have their own self-interest or good to pursue? Do we include earthworms, squids, spiders, and even bacteria in our deliberations? If we say yes, then who speaks for them? How do we assess and promote their "good" or fulfillment?

Callicott's counterargument to biocentric individualism was the contention that constructing an environmental ethic around individuals, or even species, is bound to get bogged down with those sorts of questions. Perhaps many people would find it easy to imagine that the great apes (chimpanzees, and maybe gorillas and orangutans and their biological kin)

* Just as we refer to a herd of antelope, or a murder of crows, we refer to a mischief of magpies.

have some sort of emotional sensibility. And some would find it plausible to extend such moral sensibilities to canids (dogs), felines (cats), or even corvids (ravens, crows, and magpies). But in any case, people would still get mired in questions about sentience, consciousness, and accountability of those creatures. If more kinds of beings count, how do we adjudicate conflicts between their interests? And what of conflicts that pit members of a species (i.e., a destructive nonnative species) against the healthy functioning of an ecosystem? Squids and octopi, for example, are neurologically more complex than, say, spiders. And both are more complex than earthworms and the rest of the *Annelida* phylum. Are there differing obligations depending on how complex, or advanced they are? Or rather than complexity, perhaps we should think about an individual organism's suitability to its habitat? Surely surviving in any habitat for thousands if not tens of thousands of years is a testament to an organisms' fitness. By such measures, things like sharks, crocodiles, and cockroaches have maintained very successful populations over millions of years. Sea sponges, earthworms, and some other creatures, judged by such criteria, are also hugely successful.

Callicott's approach grounded in the Land Ethic argued that the health and well-being of ecological systems was more important than individual creatures. By such reasoning, one might conclude, in some cases, that certain animals might even be removed from ecosystems if they threatened other animal or plant regimes that, although fewer in number, were more important for the resilience of the ecosystem. Rather than deciding what each of those creatures' desires are (how can we know what's "good" for a bacteria, a bat, a bear, or a bonobo?), Callicott observes, it is philosophically simpler to imagine the unit of moral considerability not as not an individual creature, but the ecosystem itself. In his famous article "Animal Liberation: A Triangular Affair," Callicott (1980) suggested that one might even imagine that a species such as a grizzly bear, or a wolf was more important than a human. Bears and wolves are keystone species, those that exact profound effects across trophic levels.[5] And because bears (especially grizzlies) and wolves are much rarer than humans they have more ecological value than humans and could therefore be imagined as more ethically important than humans. Later Callicott considered that article a source of some embarrassment, even if it is one of the most republished articles on environmental ethics. He attributed any overstatements he may have made in that article to his earnest desire to provide a positive disruption to a field that needed it, and to a logical misstep in his interpretation of Leopold's Land Ethic, the technicalities of which are outside the scope of our aims here. Callicott brought animal rights and environmental ethics together again in subsequent publications (Callicott 1984, 1988), and added an "introductory palinode" on his website which discussed the development of his thinking on the topic (Callicott 1980, 1988, 1994).[6]

Critiques of Non-anthropocentrism and New Research Avenues

Callicott'contributions to scholarly conversations about the "great wilderness debate" were important to environmental history (Callicott & Nelson 1998). It all started with historian William Cronin's famous chapter "The Trouble with Wilderness; or, Getting Back to the Wrong Nature" (Cronin 1995), which was a powerful critique of the received wilderness idea that stretched back to the late 1800s. The wilderness that Americans had celebrated as one of the unique features of their new nation was in reality only wilderness if it was devoid of human interference. That meant that wilderness was created by removing humans from the picture. In most cases, this required the forced removal of Native tribes. More importantly, said Cronin, wilderness was a social construction—it was relative to certain people's interests. If "wild" nature requires no human interference, how is that reconciled with the recognition that humans are animals, too? The feedback from Cronin's piece was swift and strong. He was criticized by several for removing the idea of wilderness to a postmodern fantasy world concocted through discourse. They rejected Cronin's claim that there was no such thing as a "true" pristine wilderness, only contestations over what that term meant, and whose ends it served. This was good news for extractive industrialists, who deployed the notion that wilderness was just a human construction in public relations battles over landscapes and resources.

The first issue of the journal *Environmental History* featured a reprint of Cronin's chapter with commentary from some of his colleagues, as well as Cronin's response. There Cronin lamented the accusation that his work relativizing the concept of wilderness provided the ammunition for extractionists and developers. Cronin declared in his response that he, too, worshipped in the "church of nature," and that such accusations reflected neither his intentions nor their desired effects. Many conservation biologists especially disliked the deconstruction of the concept of nature.

For Callicott, the concept of wilderness as the dispossession of the land and the removal of Native peoples was unacceptable and unnecessary when there were newer approaches for valuing the ecosystem services. The concept of sustainability, or sustainable development, emerged as a supposedly more holistic approach to preserving the natural world, often in cooperation with humans (World Commission on Environment and Development (WCED) 1972). Sustainable development posited that there were three legs to the sustainability "stool," the economic, the social, and the ecological.[7] A sustainable society would require the careful navigation and fulfillment of all these areas of human development. Callicott and others argued that this might be a productive way to approach nature preservation.

Some environmental ethical conversations took the form of argumentation related to intrinsic value or inherent worth. I use them somewhat interchangeably here, as some philosophers do, but others have specified key differences between them.[8] Paul Taylor's classic work of environmental ethics *Respect for Nature* (1986) was an attempt to systematize environmental ethics into a deontological form—that is, he advocated for a rule, or duty-based system. Some creatures were "teleological center[s] of life," which had their own interests and purposively sought toward their own good. Thus, they had intrinsic worth and demanded certain obligations of humans. Taylor's project was an attempt to de-center humans in ethical deliberations and to argue on the strength of objective known principles that humans were just one of many morally relevant creatures. His four corollaries were:

1 Humans are members of the community of life and membership terms are the same for all human and non-human members ("members of the community of life" is how Leopold often referred to those occupying the landscape)
2 Ecosystems in totality are interconnected and interdependent (an insight from ecology)
3 The individual organism is a teleological center of life and pursues its own good in its own way (inherent worth)
4 Realizing 1, 2, and 3, the idea that humans are superior to other creatures is groundless and must be rejected (Taylor 1986: 99–100).

If other species or natural systems are morally considerable, then they must have some sort of intrinsic value, or inherent worth.[9]

Others such as the philosopher Holmes Rolston III (e.g., 1999) argued for the objective intrinsic value (his terms) of nature. Nature's intrinsic value was illustrated by its tendency toward complexity and the "countercurrent" of energy that manifests in complexification despite entropy. Callicott argued that Rolston's approach inappropriately advocated for divine action as inherent to the unfolding of biology. Callicott approached the notion of intrinsic value in a different way: he referred often to the New Ecology and the New Physics. Some explications of them will be useful.

The New Ecology built, out of the deterministic and mechanistic ecological sciences, a concept of a living nature animated by systems theory and other advances.[10] According to Callicott the idea that "the New Physics and the New Ecology should be conceptually complementary and convergent" was the product of ecologist Arthur Tansley. It was Tansley, Callicott claimed, who pledged to take "ecology as a science out of a qualitative, descriptive stage, with anthropomorphic and mystic overtones, and transform it into a value-free, exact quantitative science" (Callicott 1985: 56). Tansley based his model of ecosystems on new advances in physics. While this model increased scientific productivity, it also raised questions about where and to whom the benefits of increased productivity accrued. While Tansley imagined purely objective and value-free sciences, there were others who still advanced an organicist scientific perspective. As Callicott put it,

> just as the philosophical interpretation of the New Physics, the Copenhagen Interpretation and its variation and alternatives, is quite another thing from its economic and military applications—from T.V. to laser weaponry—so the philosophical interpretation of the new ecology is quite another thing from its agronomic and managerial applications—from Ducks Unlimited to the green revolution.
>
> *(Callicott 1986: 57)*

In contrast to Tansley, Aldo Leopold patterned his own conceptions of ecosystems on the work of Pytor Ouspensky (1878-1947), a Theosophist, occultist, and author who was unabashed about making ecology a mystical pursuit (Ouspensky 1949 [1920]). Ouspensky's vitalistic concept of dynamic processes unfolding in time appeared in Leopold's Land Ethic, and Callicott noted this. The New Ecology was holistic, and it had not just scientific, but also moral implications. At the same time, others had already started to investigate the philosophical and theological implications of the New Physics.

One early articulation of the mystical side of the New Physics was Fritjof Capra's *The Dao of Physics* (1975), which held up interpretations of quantum theory and Eastern mysticism as roughly parallel ways of imagining a world that was deeply enfolded and "wholistic," a term coined by the physicist David Bohm (1981, 1993). Another perspective emerged from the theologian David Ray Griffin and physicist David Bohm's *Mind in Nature* (1977), which suggested that human minds were the product of, and could only be properly imagined within, the context of an evolving creation.[11] Apparently unaware of, but closely paralleling Capra, Griffin's, and Bohm's positions, Callicott posited that quantum theory could provide a morally instructive tool for humans (Callicott 1985). The fuzzy boundaries between entities at the quantum level meant that certain qualities attributed to individuals could be considered secondary qualities, rather than characteristics of discreet, bounded individuals. Callicott argued that the holistic picture of the cosmos provided by the New Physics (Davies 1990, 1989) could prove instructive for people. Callicott believed that if only people could understand that the world did not exist as a series of discreet entities, and that all these fuzzily bounded entities were engaged in perpetual dynamic relationships, humans might embrace the interconnectedness of all things, and people could grow in their respect for the natural world.

Here, then, is the rub. Non-anthropocentric positions are typically *monistic*—that is, they advance a biocentric or ecocentric ethic which is either deontological (rule and duty-based) or utilitarian (greatest good for the greatest number). But when it came to real-world conundrums, biocentric and ecocentric approaches presented problems. As noted above, questions about which creatures count, and *who* decides, is a difficult calculus. The philosopher Bryan Norton argued (1994, 1995), that in many cases (although not all) non-anthropocentric and anthropocentric ethical orientations might well agree on the proper course of action. As a pragmatist, Norton argued:

> As applied philosophers, they [non-anthropocentrists] hope to resolve environmental problems by throwing fully formed theories and principles over the edge of the ivory tower, to be used as intellectual armaments by the currently outgunned environmental activists, to aid them against the economic philistines in the political street wars that determine the fate of natural environments. This heroic version of applied philosophy's role in the policy process will only be realized if environmental ethicists, laboring in the tower, can agree on which theory to throw down to the street fighters.
>
> *(Norton 1995: 345)*

Norton's book *Toward Unity Among Environmentalists* (1994) advanced this argument in a book-length treatment, but the question implicit in his thesis was important: why aim for a non-anthropocentric ethic if a pluralistic ethical approach gives environmentalists much of what they want, still leaving room for non-anthropocentrists to push the ethical envelope?

Paul Taylor, Callicott, and others assumed that there was one ethic that should guide all decisions related to environmental issues—they were monists. Others like Norton proposed pluralistic ethical theories which they hoped might not fall into the difficulties of advancing monistic non-anthrpocentric ideologies. For instance, Norton argued (1994, 1995) the ecocentric interpretation of Leopold's Land Ethic, in the context of his own work in the Forest Service, was suspect. According to Norton, Leopold was, rather, a consummate pragmatist, and even if he had a non-anthropocentric heart, he had the practicality and the vocational commitments of a forester (Norton 2005).

The pluralistic ethic championed by Norton and other well-known philosophers accounted for several different ontologies and epistemologies, but still retained a focus on environmental well-being. One critique of monistic arguments for inherent worth is that they were derived from Western rationality and presuppositions. Even if scholars attempted to advance a holistic ethic, they did so from a philosophical foundation largely grounded in Western (often Greek) philosophical notions. And from those fundamental assumptions extended ethical consideration to other-than-human creatures (or, as we have seen, even ecosystems). Pluralistic approaches had the advantage of being able to draw on multiple ontologies and epistemologies to formulate moral norms. The cost of such norms was that they could not be generalized—they were instead contextual, specific to particular places, times, and communities of practice.

Aside from being mutli-perspectival, is there anything else to be gained by adopting a pluralistic ethic instead of the non-anthropocentric ethical stance advocated by Callicott and others? Well, a localized ethic does have its advantages. Non-anthropocentric ethics, like the ones recounted above, are what Jim Cheney and Anthony Weston (1999) once called *epistemology-based ethics*—this includes most Western ethical formulations. An epistemology-based ethics depends on several presuppositions. First, ethical action is a response to knowing the world. To put it differently, knowledge about the world comes first, and then we might prioritize specific subsequent actions. Second, such models assume that the world was readily knowable, or at least that it was possible to know enough to make ethical decisions and distinctions in the world. Third—and this one is true for both biocentric and ecocentric ethical formulations—building an environmental ethic was an incrementalist and extentionist business. Once we know the world, then reason might be stretched to include other-than-human creatures in our ethical deliberations. In contrast, Cheney and Weston offered an *ethics-based epistemology* (1999), where the ethical calculus was grounded in relationship and relevance, with more formalized "rational" assessments to follow. People, places, and things gained significance for people through interactions with them. Things central to human lives required engaging in relationships with those things. From that relationship, moral bonds form. It is from an ethical point of view, from that starting point, that notions of right or wrong actions emerge.

Comparative Environmental Religions and Ethics

Frustrated by the inadequacy of secular philosophical ethics in generating a useful environmental ethic, Callicott engaged in a groundbreaking and broadly conceived comparison of environmental ethics "from the Mediterranean basin to the Australian outback" (Callicott 1994). In his comprehensive cataloging, Callicott suggested that the religious beliefs and practices of the people he surveyed could foster a common environmental ethics if those groups would only deemphasize their supernatural claims.[12] As noted, he tethered these various cultural manifestations of respect for nature to the then emerging postmodern scientific paradigm (the New Ecology and the New Physics), amenable to a host of different worldviews.[13] With a similar approach the philosopher Max Oelschlager also imagined that religious beliefs were crucial elements for fixing the environmental crisis. He argued that the "Judeo-Christian" tradition was a global phenomenon which exerted its influence even beyond its practitioners through the engines of global capitalism (Oelschlager 1994). The dualism so deeply ingrained in the Western imagination had to be chastened before Westerners would be able to properly "care for creation." If those religious stories were imagined as guiding narratives rather than literal truths, they may well contain the seeds of a viable environmental ethic.

Both approaches explicitly addressed the ideational component of environmental ethics. That is, both Oelschlager and Callicott suggested that once people got their ideas about nature right (once they understood the holistic nature of things), their behaviors would follow. Both Callicott and Oelschlager, then, turned toward religious ideas as a potential corrective to traditional philosophical ethics. Interestingly both also thought that their contribution to creating a sustainable world depended on purging religions of their supernaturalistic and literalist ideas.

For a helpful contrast, Fikret Berkes (*Sacred Ecology* 1999) provided a detailed study of the James Bay Cree, an Indigenous population in Canada. He drew on both native stories and Western science, both etic and emic descriptions of behavior, to note where those two approaches cohered.[14] He detailed how human behavior adapted over time to resource overexploitation or shortage. Berkes noted that the ethical components of the Cree worldview were coded in stories about practices, not abstract beliefs. In the end, Berkes suggested that Indigenous epistemologies should be allowed a significant role in resource use and management, since they have proven adaptable, and generally sustainable over time. These alternative epistemologies could be joined with postmodern scientific methodologies under the rubric of "sustainability." Berkes's story has global implications, if indeed several Indigenous epistemologies might be fruitfully combined with Western scientific observation (Hukkinen 2008; Callicott 2011). Berkes and others acknowledged that local populations' religious complexes could be fruitfully combined with postmodern science, (Berkes et al. 1998, 2003; Berkes 1999, 2018). Rather than downplaying the religious dimensions of their practices, the Cree did not perceive that the category of "religion" was something that could (or should) be separated from their everyday lifeways.

Investing in comparative approaches to environmental ethics was potentially ameliorative as globalization caused the acceleration of temporal and spatial interactions. International bodies like the United Nations

Development Program (UNDP) and the United Nations Environmental Program (UNEP) admitted Indigenous and traditional lifeways and epistemologies into the logic of resource use and management. It was, however, a long process to get Indigenous voices around the negotiating table. Although not part of the official conference, Indigenous peoples mustered a collective voice at the Rio Earth Summit (Hart 2005) and subsequent conferences.[15] These events, and the rubric of "sustainability," all suggested that traditionally marginalized voices should grow more important in how development proceeded, and how development itself was defined. Because of these shifts, it was vitally important that alternative epistemologies were investigated from both etic (outsider) and emic (insider) perspectives. Comparative projects that described consonances and dissonances between how these various ethics and epistemologies were conceived and embodied provided some helpful pathways for comparative environmental ethics. Their direct implications for policy formulation made these comparative projects one of the most practical contributions to the transformation of human interactions with the non-human world.

So, there were non-anthropocentric (often monist) environmental ethical approaches which might still be consonant with pluralist approaches *in practice*. This was the crux of Norton's argument (Norton 1994). And yet, Norton *rejected* non-anthropocentric approaches, rather than acknowledging that monist ethical formulations might have their place in policy deliberations also. Much international law and policy in the twenty-first century was based on the notion that humans have some fundamental "worth" or "dignity," and should be accorded certain rights. In some cases, the intrinsic value and worth of ecosystems was advanced in international policy documents. For instance, the United Nations publication *Man in the Biosphere* advanced a strongly biocentric vision focused on the intrinsic value of other-than-human species and the inherent worth of natural systems (United Nations 1986). Callicott and others published, in scientifically oriented journals, advocacy of biocentric and ecocentric themes and their importance to the policy formation process. In the twenty-first century, there has been a trend toward protecting species, and ecosystems, and landscapes or seascapes using the language of rights and duties (Epstein et al. 2023).[16] Such legal actions are grounded in non-anthropocentric ethical perspectives. Many of the policies related to rights-based argumentation were light green. But when the places and natural spaces were protected because they were sacred sites, they exposed dark green roots. One possible avenue for future research might be a deeper investigation of whether and to what extent dark green sensibilities inform such policy measures.

Importantly, however, Callicott also recommended that cross-cultural comparisons could prove useful in the quest for a global-scale ethic. First, it was necessary to advance an ethic that did not depend on some of the individualist moral philosophies detailed above. Recall that in deontological (rule-based) and utilitarian (use-based) schemes, the unit of moral considerability (who or what "counts" in our ethical deliberations) was vexed with questions regarding individual sentience of organisms, or about maximizing the good of the greatest numbers of individuals. Callicott's work made it clear that more radical work was required by philosophers in the twenty-first century if humans were to finally find a way to cohabitate with other creatures. It required, he argued, a new epistemological framework, which he assembled from the natural philosophies of David Hume (1711–1776), Charles Darwin (1809–1882), and Aldo Leopold (1887–1948). Hume's ideas about moral sentiments were "biologized" by Charles Darwin by linking morality and pro-social behavior. Aldo Leopold writing a century later than Darwin "ecologized" this narrative about the evolutionary-ecological roots of morality. In other words, with a new epistemology grounded in biologized conceptions of functionally interconnected parts came a new ontology, one where ecological collectives were the fundamental unit of moral considerability. Through an ecological lens, the human individual is a fabrication.

Worldview Remediation Is an Environmental Religion

Leopold's antagonism toward biblical religions in particular, and Abrahamic ones in general, was really an attack on a worldview that had outlived its usefulness. The religious worldview, it became clear, was leading humankind down a path which would end, ultimately, in its own demise. A different worldview took shape in the middle of the twentieth century as the discipline of ecology flowered (Worster 1979) into what Callicott termed the *evolutionary-ecological* worldview (Callicott 2011).

Callicott traced the concept of worldview back to the eighteenth-century philosopher Immanuel Kant (1724–1804). Kant imagined the worldview derived from the way the world appeared through human sensory capacities. The phenomenal experience of the world was always constrained by the limitations of human faculties. Was there a *really real* world outside of what our innate faculties could detect? Kant believed there was, and he referred to it as the *numina,* things as they really were before they were interpreted by human brains. Human brains had innate cognitive capacities to categorize things, and it was only through these faculties that the objective reality, the *numina*, could be obliquely encountered. In the twentieth century, some scholars of religions exercised the notion of worldviews to describe various religious social formations (Smart 1987, 1996; Tucker & Grim 1994; Taves & Kinsella 2018). More recent manifestations of worldviews have deconstructed the concept to include ontological, epistemological, axiological, praxeological, and cosmological dimensions (Taves & Kinsella 2018). Those elements of worldviews, and the ways in which social groups prioritize them differently, might be a productive way for the concept to do comparative work (Johnston 2018).

Teasing apart the ontological, epistemological, axiological, praxeological, and cosmological elements and analyzing how different groups prioritize those elements is a good first step toward worldview remediation. For Callicott, armed with an evolutionary-ecological worldview, remediation required revisiting and rejecting outdated religious paradigms. Callicott emphasized that although scientific descriptions of the world were superior to pre-ecological worldviews (before the science of ecology emerged as a discipline), they could still be described in narrative ways that added emotional weight to ethical reasoning. As Callicott put it, "Exponents of the scientific worldview, such as Leopold, can effectively draw on the discourse of religion and mythology to express the scientific worldview in ways that are accessible and persuasive to a lay audience" (Callicott 2011: 517).

Leopold deployed religion resembling rhetoric to rail against the prevailing and outdated Abrahamic worldview. With a criticism that foreshadowed Lynn White, Jr.'s famous article (1967), Leopold wrote: "Abraham knew what the land was for: it was to drip milk and honey into Abraham's mouth. At the present moment, the assurance with which we regard this assumption is inverse to our education" (Leopold 1949: 204–205; quoted in Callicott 2011: 511; Callicott 2013). Re-envisioning humans *as a part of* the community of life rather than its supervisor, Leopold's vision of the evolutionary-ecological worldview provided a template for worldview remediation in the 21st century, and Callicott was its primary champion (Callicott 2013, 2011).

Engaged anthropology, conservation biology, and restoration ecology were all problem-based disciplines that emerged in the latter half of the twentieth century (Wright 1988; Takacs 1996; Jordan & Lubick 2011; Johnston 2016;). Each of these were sub-disciplines that reinvigorated the biological sciences by returning to some of the moral and often spiritual dimensions of scientific inquiry. William Jordan, III, the co-founder of the Society for Ecological Restoration and the affiliated journal's first editor, portrayed the science of restoration ecology as a way of doing penance for humans' destructive impacts (Katz 1992, 2012; Elliott 1997; Jordan & Lubick 2011; Light & Higgs 1996). These were attempts to enact worldview remediation. In the twenty-first century, it is common to see spiritualized versions of the evolutionary-ecological worldview on display, especially in cultural contestations over land use (see Chapter 5, for instance). The paradigm shift from a biblically inspired human exceptionalism to one where humans are plain citizens of the ecosystem is still underway. Callicott blazed the philosophical path for it by resuscitating Leopold's notion of worldview remediation. As readers we can hope that the change proceeds quickly, given the acuteness of environmental problems. To the extent that a new evolutionary-ecological worldview challenges outdated religiously inspired anthropologies, it has become an environmental religion.

QUESTIONS FOR DISCUSSION:

- In your own words, define what it means to have an *anthropocentric*, and a *non-anthropocentric* ethical stance? Is one more convincing, or appropriate in your estimation?
- What is the difference between utilitarian ethics, and deontological ethics? Give an exemplar of each sort of ethical reasoning.
- Do you think it is necessary to purge religions of their supernatural elements for them to mesh with science?
- Do you think that Callicott's notion of worldview remediation is useful? Why or why not?

Notes

1 Personal conversation, Winston-Salem (2015).

2 I take this from the title of a book chapter by Gavin Van Horn (Bauman et al. 2011).

3 Jeremy Benthan (1748–1832) is typically credited with fully developing a utilitarian ethical formulation.

4 Personal conversation, 2008.

5 Trophic levels are elements of the food web. Producers, organisms that transform energy from the sun, are on the bottom trophic layer. The second level may be populated by herbivores who feed on the plants for sustenance. There may be carnivores that eat the herbivores, representing another level, etc.

6 See https://jbcallicott.weebly.com/introductory-palinode.html, accessed 11 August 2022. In short, because the term *when* does not have technical meaning in philosophical logic it can mean either *if* or *if-and-only-if*. Callicott's argument stands if the former is assumed, but not the latter. His correction is explained in more detail on his webpage.

7 This metaphor is rarely, if at all, deployed in sustainability oriented social movements at this writing.

8 Paul Taylor differentiates between intrinsic value (accorded to an event or condition), inherent value (which is placed on an object or a place), and inherent worth (appropriate only for creatures which have a good of their own) (Taylor 1986: 73–74). For any entity X to have inherent value, "a state of affairs in which the good of X is realized better than an otherwise similar state of affairs in which it is not realized…, (a) independently of X's being valued, either intrinsically or instrumentally, by some human valuer, and (b) independently of X's being in fact useful in further the ends of a conscious being or in furthering the realization of some other being's good, human or non-human, conscious or unconscious" (Taylor 1986: 75).

9 The New Physics emerged out of the seemingly illogical world of quantum mechanics in the early part of the twentieth century. The Copenhagen

interpretation of quantum mechanics was so called because some of the most important contributors did their research at the University of Copenhagen. One of the moss most important elements of the theory was that there was an inescapable indeterminacy at the quantum level. The Uncertainty Principle (part of the Copenhagen model) suggested that either position or velocity of a subatomic entity might be known with accuracy, but an accurate description of both was impossible. The act of measurement affected the system in significant ways. If an observer wishes to know an object's position, its velocity is affected. Likewise, if an experimenter aims to measure velocity, then it is impossible to discern the exact location of a particle. Also important to the Copenhagen Interpretation was the Principle of Complementarity, which stated that, depending on the experimental apparatus, subatomic entities might appear as either a wave, or as a particle (see Herbert 1985; Davies 1990a, 1990b; Chiao 2011 for further explanation).

10 General systems theory is an approach that views biological entities as parts of a whole, or system, where the individual cannot be abstracted from the holistic view of the system (von Bertalanffy 1972: 10).

11 Griffin was a theologian at Claremont University, and Bohm was a physicist who advanced what Callicott refers to as the New Physics.

12 Callicott's book is not limited to the world religions, but also includes analysis of Indigenous cosmologies and ethics. His survey of environmental ethics "from the Mediterranean Basin to the Australian outback" was cited and affirmed in the Series Foreword for volumes on the World Religions and Ecology, published by Harvard's Center for the Study of World Religions.

13 The term postmodern can mean different things when deployed in different circumstances, and disciplinary orbits. In general, it can be said that most postmodern positions questioned the validity of foundationalism, the idea that there were objective realities in the world independent from human encounter and experience.

14 Emic is an anthropological term which refers to an "insider" viewpoint, one who is a practitioner of a particular worldview. An etic approach refers to the perspective of one who is not a participant in a group.

15 See Taylor (2010) for an account of the dark green religion embedded in the 2002 Johannesburg conference

16 See https://www.science.org/doi/10.1126/science.adf4155; https://sdgs.un.org/partnerships/rights-nature-catalyst-implementation-sustainable-development-agenda-water; https://news.climate.columbia.edu/2021/04/22/rights-of-nature-lawsuits/, all accessed 9 August 2023.

9

VANDANA SHIVA

"Stolen Seeds, Simple Needs" (Physics, Agroecology)

The Green Revolution and a Monoculture of the Mind

Vandana Shiva is a criminal. Her non-governmental organization, Navdanya, focused on saving seeds, conserving biodiversity, and promoting agroecology. Doing this work meant that Navdanya violated patent laws, and as such was a subversive and technically illegal organization with 122 seed banks in 18 states of India and Bhutan.[1] In India and elsewhere a legal and national capitulation to multinational corporations made traditional farming techniques and markets obsolete. The price of entry into the global grain market was an acceptance of servitude to corporations' vertically integrated production processes. But how did agricultural seed and chemical manufacturers conjure the ability to set international laws and standards which drive local farmers out of business?

To answer such a question, it is necessary to cast our eyes back to the close of World War II (WWII) and the creation of the Bretton Woods institutions. They are so-called because they were agreed upon at a meeting of WWII victors in Bretton Woods, Maine. It was the incubator for new international corporate and governmental enforcement mechanisms. The military-industrial complex and some corporations, which have all the rights of people but none of the liabilities, drove the policy-making process. At the close of WWII, the social-democratic response to the global war rested in the hands of the new United Nations organizations. The financial and development sectors, however, were the provenance of the Bretton Woods organizations: the World Bank (WB), the General Agreement on Tariffs and Trade (GATT) (later renamed the World Trade Organization (WTO)), and the International Monetary Fund (IMF). Many, including U.S. President Harry Truman, concluded

DOI: 10.4324/9781003364801-10

that WWII was an inevitable consequence of the crippling sanctions levied against Germany and her allies following WWI. Crushing poverty and inflation which gripped Germany after the first war was imagined as a major contributor to the social and economic upheaval that allowed Adolph Hitler to rise to power. An intention to engage in a different strategy at the end of WWII resulted in a post-war plan for *development*. If the underlying cause which led to the rise of fascist dictators in Europe—poverty—was cured, then the world might have a chance at a *Pax Americana*, a global order presided over by the United States and their allies. Development, and later sustainable development became an ideal for the world. Typically, this meant that the WB and IMF lent money to countries in the global South on the condition that they would shift their mode of production from subsistence farming and diversified economies into making and selling commodities for a global economy. The reconstruction of national economies based on exported commodities was called *structural adjustment*. Imagined largely by economists from the University of Chicago, and largely implemented by the Bretton Woods institutions, structural adjustments were made to shape the economic arrangements in the global South to complement the existing industrial capacity of the global North.[2] Structural adjustment impacted not just economics but also the political, ecological, and social fabric of traditional communities. Indigenous populations, already on the margins, were sometimes referred to as the "Fourth World," populations within the global South that were marginalized from the global economy, but also politically isolated and impotent in their own countries (Wright 1988).

Pricing of global commodities undermined traditional lifeways so that in some cases imported foreign goods from far away became cheaper than those produced locally. Community-scale livestock and crop producers, and other local commercial enterprises died off. As the economy became truly global, biodiversity and cultural diversity eroded. The logic behind this brand of capitalism ultimately led the world into monocultural monotony: a handful of genetically modified crop (GMO) varieties, fewer species of livestock, and a global human culture that ate at the whim of vertically integrated agricultural corporations.

In her *Stolen Harvest* (2000), Shiva provided illustrative examples of local trades that went extinct under the new economic regimes of *development* and *sustainable development*. In the Indian context, she highlighted the disappearance of plant and seed oils unique to each state, and the explosion in aquaculture that destroyed place-based wisdom and community cohesion. The result of structural adjustment was "the flooding of domestic markets with artificially cheap imports stealing local markets and livelihoods from local farmers and local food processors. The expansion of global markets [was] taking place by extinguishing local economies and cultures" (Shiva 2002: 22). As the industrialized world's agricultural producers navigated the 20th century, they conglomerated into six large companies that shaped the way the world food system worked.[3] In the process, Monsanto produced a genetically modified species of soybean that was resistant to weed killer. Monsanto owned the patents on those soy plants, but also on the technologies used to support their growth: herbicides, pesticides, and mechanized sowing processes (Shiva 2002: 29). Life itself was patented and declared the property of the corporations that "invented" modified life forms.

The Green Revolution was one name given to the coalescence of agribusiness and development which emerged and gained steam through the latter half of the twentieth century. Often credit for this development is traced to the agricultural scientist Norman Borlaug (1914–2009) who pioneered some of the first widely distributed genetic plant crops, including a drought-resistant hybrid rice, and a dwarf wheat variety that dramatically increased yields in the developing world. Shiva categorized agricultural technologies into three periods of industrialization: "The first industrialisation was based on the mechanisation of work, with focus on the textile industry. The second industrialisation was based on the chemicalisation of processes in agriculture and other sectors, and the emerging third industrialisation is based on the engineering of life processes" (Shiva 2016: 235). Patenting and privatizing life and life processes was a contested issue. On the one hand, there were people who might view privatized gene modification as tampering with nature, disrupting its "purity," or interfering with "natural" processes. On the other hand, were those who argued that hybridization and privatization are themselves natural processes developed and executed by natural creatures (humans), and given this capacity humans are obliged to utilize it to control, shape, or liquidate nature for human utility. These are questions for environmental ethics and are explored more fully in Chapter 8.

Periods of industrialization dramatically impacted all facets of the global South erasing the diversity of ways in which people historically created food systems. To create new strains of crops, Western scientific methods displaced the empirical "trial and error" method used over generations. It was a global simplification of epistemologies: "Development in this view [was] taken as synonymous with the introduction of Western Science and Technology in non-Western contexts. The magical identity is development = modernisation = Westernisation" (Shiva 2016: 233). Biotechnology thus generated and preserved global human inequalities; a globalization of one (Western) epistemology, creating global monocultures of the mind (Shiva 1993, 1994). Shiva observed that biotechnology was thus an instrument for dispossessing farmers of seed in an ongoing a high-cost means of production. The relocation of seed production from the farm to the corporate lab "relocate[d] power and value between the North and South; and between corporations and farmers" (Shiva 2016: 245).

Shiva used the mustard plant as an example, a generations-old source of cooking oil which suddenly became essentially illegal to grow, and nearly impossible to acquire. This was because Monsanto, one of the world's largest agricultural-chemical manufacturers, patented the Indian mustard plant (*India brassica*). Then, they removed it from the market. If it were to be reintroduced to the world in the future, it would exist only as a patented product, a proprietary form of life owned by a corporation. A genetically modified product which required high inputs of fertilizers and pesticides, which where also owned by the seed companies.

Shiva recounted other similar measures that eliminated small producers in favor of mass production. Corporate food giants simplified biological diversity, so crop yields were predictable and controllable. Shrimp farming, an industry that was booming in the early 2000s, led to the clearing and commercial development of mangrove swamps. Mangroves provide important ecosystem services such as creating bird and aquatic animal habitats and increasing coastal resilience in the face of disturbance (i.e., typhoons). Although the public corporate rationale for increasing yields was combating hunger in the global South, the fact was that most of that shrimp was consumed by people in the United States and Europe—it was a delicacy that the farmers themselves could not afford to eat. Structural adjustment, then, did little to help the farmers who traditionally depended on those mangroves and other associated species (Piaget-Clark & Shiva 2004). In the capitalist model those social and ecological services were treated as externalities, things that were not figured in when assessing the cost of food production. As a resource became scarcer its economic value increased—there was literally an incentive to cut down the last tree, kill off the last rhino, and plow the most marginal lands if it resulted in economic growth. Environmental decline sped up, and corporations maximized their ability to produce increasingly rare and thus expensive products (whether oil, gold, uranium, coal, or other materials).

The new agricultural products increased yields because they integrated the seed, pesticide and herbicide treatments, and harvesting technologies, ultimately maximizing efficiency, but only for those who had access to the resources to procure large-scale harvesting devices and other technologies.

> The important feature of these new [seed] varieties," Shiva noted, "[was] not that they [were] particularly productive in themselves but that they absorb[ed] three- or four-times higher doses of fertilizer than the traditional varieties and convert[ed] it into grain, provided proportionately heavy and frequent irrigation applications [were] also available. They also [had] high susceptibility to insect and pest attacks,
>
> *(Shiva 1993: 247)*

Importantly, Shiva connected the dots between food, the capitalist mode of production, and marginalized populations. While India was her home, Shiva's work shed light on places all over the world where similar contestations occurred. More specifically Shiva provided a voice for the poor, especially for poor women worldwide who suffered the additional injustice of being born into a patriarchal world. It was this care for food systems and the rural poor which guided her contributions to another academic discourse, ecofeminism.

Ecofeminism and Globalization

"As women in various movements—ecology, peace, feminist and especially health—rediscovered the interdependence and connectedness of everything, they also rediscovered what was called the spiritual dimension of life," according to Shiva and the German sociologist Maria Miles (Miles & Shiva 1993: 16). In fact, the seemingly burgeoning interest in non-institutionalized spiritual practice "[was] a manifestation of Western patriarchal capitalist civilization's deep crisis" (Miles & Shiva 1993: 19). In the latter half of the twentieth century and early twenty-first century, many divested from the mainstream institutionalized religions, often in favor of a spiritually eclectic combination which often included American Indian, or Eastern (mostly Hindu and Buddhist) mysticism, rites, and relics. Recall that Durkheim referred to this societal disease as anomie, a disquieting perception that capitalist markets were hollow, empty in the middle. Like the façade of a building, economic activity put a pretty public face on an interior that was vacuous and valueless. Removed by more than 100 years from Durkheim, Shiva witnessed the maturation of the anomie noted by scholars following the Industrial Revolution (Weber 2005 [1930]).

Shiva's theoretical approach was valuable and reminiscent of scholars of religions who advocated a critical approach to religious studies. When industrial capitalism spread across the globe, "the dominant discourse the 'global' is the political space in which the dominant local seeks global control and frees itself of any local and national control. But, contrary to what it suggests, the global does not represent universal human interest but a particular local and parochial interest which has been globalized through its reach and control" (Miles & Shiva 1993: 9). Others such as the scholar of religions Russell McCutcheon argued that disguising ethnocentric idiosyncrasies as universals was typified in essentialist definitions of religion. Essentialist definitions target a non-empirical *thing*, some essence, that is universal among all religions.[4] The dominant local (the supposedly universal "West") was detached from its local spawning grounds and imagined as something global—a manufactured and superficial common goal. In this case, the history of Western scientific thought naturalized ethnocentric perspectives, making Western notions of both *nature* and *religion* appear universal. Making the local and contingent into *something that goes without saying* was the strategy of hegemonic powers during the colonialization and domination of people and other resources. As McCutcheon put it, "we see the common strategy of employing the comparative method to identify non-empirical commonality, such that certain types of all too obvious difference are understood to be nonessential features of contingent history. This is characteristic of those who study the world's religions" (McCutcheon 2007, 2013; McCutcheon 2019: 24).[5] To unpack this claim, McCutcheon meant that cross-cultural comparison typically posited some commonality, but that this commonality was the product of social construction rather than observable evidence. So, for instance, when scholars said that the essence of religion was a private, interior, subjective experience not ultimately sharable with anyone else, and said that all humans have *that* experience of religion, they were making a claim for which there was no data. We see or imagine the experiences of others, but we cannot experience things *for* them—so religion is ultimately a private sensation. The claim that religious people are all traveling up the same mountain, toward the same end goal (salvation), just using culturally specific language and practices, is a political claim for which there is no evidence.

A domination-centered mode of globalization and Western scientific ideas were not, Shiva contended, mere accidents of history. Rather they were produced through a unique way of life first observable following the second phase of urbanization (first millennium BCE) in India. To explain, imagine the first urbanizing groups emerged approximately 10,000 years ago. By the beginning of the so-called Axial Age, approximately 800 BCE, the major civilizational hubs were cities built by societies that had already been through several shifts in the mode of production.[6] The switch to increasingly sedentary agriculture spurred the establishment of those cities. It took only 5,000–7,000 years after sedentary agriculture became widespread—the geological blink of an eye—to see the emergence of cities in the archaeological record.[7] In other words, the first cities were, by definition, human settlements large and dense enough to require the importation of resources to survive. When these societies matured, they birthed the gods of the city (rather than the gods of pastoral or hunter-gatherer communities). And those gods of the city preserved urban lifeways through the specialization of labor, hierarchical decision-making, and social stratification. As Shiva and Miles argued,

> because capitalist patriarchy or 'modern' civilization is based on a cosmology and anthropology that structurally dichotomizes reality, and hierarchically opposes the two parts to each other: the one always considered superior, always thriving, and progressing at the expense of the other. Thus, nature is subordinated to [hu]man; woman to man; consumption to production; and the local to the global, and so on. Feminists have long

> criticized this dichotomy, particularly the structural division of man and nature, which is seen as analogous to that of man and woman.
>
> *(Miles & Shiva 1993: 5)*

Hierarchical social arrangements grew and were reinforced by a competitive perception of human relationships. In Mary Douglas's terms, those societies exhibited a consistently weakening grid. As Shiva and Miles put it,

> Modern economic theory sees self-interest as the impulse of all economic activity. Later, Darwin 'discovered' a similar principle in nature. Accordingly, the symbioses, the interconnections that nurture and sustain life are ignored, and both natural evolution and social dynamics are perceived as impelled by a constant struggle of the stronger against the weaker, by constant warfare. Such a worldview militates against an appreciation of the enriching potential of the diversity of life and cultures, which instead are experienced as divisive and threatening. Attempts to rejoin the atomized parts lead only to standardization and to homogenization by eliminating diversity and qualitative differences. An ecofeminist perspective propounds the need for a new cosmology and a new anthropology which recognizes that life in nature (which includes human beings) is maintained by means of co-operation, and mutual care and love.
>
> *(Miles & Shiva 1993: 6)*

Unfortunately, the inertia of patriarchal civilizations was difficult to overcome, particularly when the conditions of participation in globalization first required the acceptance of dualistic and hierarchical understandings of all life. When introduced to non-Western communities, colonialism resulted in both interpersonal and structural domination. Hierarchical and exploitative relationships spread like a contagious cultural virus with the universal promise of development, but weighed down with the baggage of class and race. With the global ambitions of development and later sustainable development,

> the victims [were] always local, for example, as [was] manifest in the aftermath of the Gulf War — a war justified by the apparently universal or global principle of justice, in the name of the 'world community,' represented by the United Nations. The world was called upon to feel responsibility for liberating Kuwait from Iraqi occupation. But it is clear that the victims of this 'liberation' are local: Iraqi and Kuwaiti women and children, the Kurds, and the Gulf region's environment.
>
> *(Miles & Shiva 1993: 9)*[8]

The freedom and democracy promised by Western powers was perceived by many in the global South as a cloaked statement of American global dominion. Meanwhile, those in the North advertised it as the promotion of democracy and inclusive politics. Desert Storm, the first engagement in Iraq, the subsequent conflict called Operation Enduring Freedom, and the later invasion of Afghanistan (the United States's longest military engagement) hid the imperialist vision behind cries of capitalism, democracy, and national security. Like Rosemary Radford Ruether, Shiva saw that the real culprits of the war on women and the environment were the tools of Empire—military imperialism and democratic absolutism fueled by the commodification of nature.

Ironically, all human societies represent just 0.01% of the biomass of the planet, yet they occupy at least 20% of the planet's Net Primary Productivity (NPP). To put it bluntly that is rather large proportion of the Earth's productive capacity commanded by just one clever primate species. Humans developed technologies that extended their reach, and their resultant impacts. Other animals use tools also, so it is not tools or technology that are, by themselves, the problem. It is just that some human tools are planet-changers, while the tools of others (magpies, beavers, bonobos, and more) are not to such an extent. For example, beavers are master builders of habitats, changing the ecosystem conditions with their creative housing projects (beaver dams). But those impacts are not global in scope. Humans' share of the Earth's productivity continues to grow while the planet experiences what some have called the Sixth Great Extinction event.[9] The drawdown of biodiversity continues at an alarming rate, even as the drawdown of carbon in the atmosphere needed to reach international agreements and prevent the most catastrophic consequences of climate change, never materialized.

Both fragmentation of habitats and intentional removal of species from food systems meant the diversity of seed stocks was whittled down to a few that were controlled by agriscience corporations. Exploiting a relatively minor point in the text of the 14th Amendment to the U.S. Constitution, corporations were able to argue that they had legal personhood.[10] The legal representation of corporations as persons has continued since the passage of the Amendment in 1866. By the 2010s cases such as *Citizens United vs. the Federal Elections Commission* (FEC), and *Burwell vs. Hobby Lobby Stores* successfully argued that business owners could decline services to consumers because a particular religious perspective was considered freedom of speech (the 1st Amendment). Corporations do not just have rights; they have a right to control and disperse (or withhold) resources.

As noted, corporations vertically consolidated their business, expanding from seeds to chemical enhancements, to harvesting innovations, in the post-WWII period. The weaponized chemicals deployed in warfare were tweaked for widespread commercial use. Further market consolidation meant the world's food supply was almost entirely dependent on a handful of transnational corporations whose motive was profit, and whose business was the commodification of nature. Just as in the other development programs financed by the Bretton Woods institutions, money and power floated to the top of the social/political ladder. The resultant peasant resistance movements and pan-Indigenous rights campaigns noted by Ruther and others in their work on liberation theology, emerged across the globe as traditional lifeways were eradicated through the economic liquidation of resources, which was counted as profit (Stavenhagen 1970; Taussig 1982). The costs of harvesting and distributing these resources were discounted as externalities. Ecofeminism was a tactic in Shiva's broader strategy to combat the violence of the Green Revolution. India was one of many places that absorbed this neo-colonial imposition of a new mode of production and its aftereffects—it was the globalization of anomie.

Chipko and Navdanya: Roots and Shoots

In 1973, Indian farmers won a small victory against an intrusive logging operation near the village of Gopeshwar, Uttar Pradesh. The following year in another logging contestation, when the village men went to town to confront the authorities, the loggers sensed an opportunity and set about to harvest the timber while the men were gone. What they encountered was a 50-year-old woman, Guara Devi, who reportedly told the loggers, "This forest is like our mother. You will have to shoot me before you can cut it down" (Craddock & Lal 2005: 300). The term *Chipko* means to cling or embrace—in this case it referred to hugging trees to prevent them being logged (Craddock & Lal 2005: 300). The movement gathered steam thanks to well-known activists Sunderlal Bahuguna (James 2013) and Chandi Prasad Bhatt (Lal 2005), who grew their approach with inspiration from Gandhi's nonviolent resistance (*cf.* Guha 1989, 1995, 1996).[11] Bahuguna, at least, drew on Vedic texts in his advocacy (Craddock & Lal 2005). Although those men were instrumental in spreading the message, the movement was nurtured by women who were the educational and activist arm of the movement. Women grew the roots of feminist social movements in India in the 20th and 21st centuries:

> The Chipko movement is commonly viewed as challenging Western notions of economic development that destroys the natural environment and further impoverishes the poor, promoting instead environmentally sound ideas of sustainable development and the empowerment and self-determination of the local people.
>
> *(Craddock and Lal 2005: 300)*

It was from those roots that more women-led movements emerged, many of them focused on food rather than tree species. But those movements combated the same villain: the liquidation of nature for the profit of international corporations and the erosion of Indigenous lifeways (Shiva 1996, 2016).

Vandana Shiva's work was one exemplar of those movements. She was the engine behind the mobilization of women all over the world who, perhaps initially, celebrated Green Revolution technology and its high yields, but later confronted the social and ecological costs. Awed at first by its bountiful yields, farmers soon realized that in betraying their traditional lifeways, they contributed to the destruction of their homes in search of profit. In her work *Staying Alive: Women, Ecology, and Survival in India* (1988), Shiva coined the term *maldevelopment* to challenge the prevailing economically driven mode of sustainable development, which unified the world economically but left Indigenous and traditional people out of the social calculus. From a traditional religious perspective, Western economic and developmental progress interrupted the positive feedback loops between the female (Shakti) and male energies (Purusha) which sustained the world. To fight back against the Western scientifically grounded, capitalist-fueled, monolithic solutions to complex and contested contexts, Shiva promoted an Earth Democracy. In such a system the inalienable rights of Mother Earth were respected by the cultivation of just human societies promoting ecological and global well-being. The Principles of Earth Democracy were:

1. All species, peoples, and cultures have intrinsic worth
2. The earth community is a democracy of all life
3. Diversity in nature and culture must be defended
4. All beings have a natural right to sustenance
5. Earth Democracy is based on living economies and economic democracy
6. Living economies are built on local economies
7. Earth Democracy is a living democracy
8. Earth Democracy is based on living cultures
9. Living cultures are life nourishing
10. Earth Democracy globalizes peace, care, and compassion (Shiva 2005: 9-11).

These principles build on one another, fostering a positive feedback loop that increases the capacity of social groups for compassion, peaceful cooperation, and diversity in democracy. In another context, Shiva elaborated on the principles of Earth Democracy:

1. The Earth is living. The Living Earth is our Mother. She is Terra Madre, Mother Earth, Gaia, Pachamama, Vasundhara...Mother Earth has rights.
2. We are all members of the One Earth Family. We are part of the Earth, and not separate from her, not her masters. We are interconnected through

the living currencies of breath, water, and nourishment. We have a duty to protect the Earth's Living Systems that provide us clean air, clean water, and clean food.
3 We are part of One Humanity on One Planet. All humans are equal. Our diversity enriches life and cannot be made the justification for inequality and injustice. Future generations have a right to enjoy the gifts of the Earth. Present generations have a duty of Earth Care to pass on the gifts of nature in their full diversity, integrity, and purity.
4 Earth Democracy is based on Living Economic, Living Democracies, and Living Culture woven through the diversity in the web of life. Each life form supports and sustains all others in mutuality and cooperation and harmony. All living beings are sentient beings and have rights. All beings are creative and intelligent".[12]

Earth Democracy is "a worldview, paradigm, and practice that was based on the recognition that everything is interconnected, the Earth was Living, the Earth had rights, that we had duties to care for the Earth, and regenerate her soil, seeds and biodiversity, and her water and food systems. Our rights flow like a spring from our duties," Shiva declared (Shiva 202: 6–7).[13] Navdanya was an educational enterprise, a women's empowerment movement, and a seed saving bank, all fed by *agroecology*, the preservation of ancient agricultural practices which provided ample subsistence to local farmers. Navdanya used biological pest controls (rather than chemical ones), created context-appropriate crop rotations, and promoted a resource commons, freely sharing and preserving knowledge through women-focused networks across India. According to their website, Navdanya reached over 750,000 farmers and created 150 seed banks in 22 Indian states.[14] Those are the shoots of women-led, agricultural communities which resist ubiquitous but malignant modified seeds.

Predictably, this made Shiva dangerous to extractive interests. The laws she broke were written by lobbyists for the enrichment of Big Agriculture. She was attacked in some reputable magazines and always in opinion pieces in the newspapers of cities where she spoke. Oftentimes the op-eds conveyed caricatures of her position, insisting that biotechnology was necessary to feed the growing global population. There was always a dramatic misunderstanding (whether intentional or not) of her mission and strategies (Entine 2014; Specter 2014; Tagliabue 2021). There was a cadre of people who preceded her everywhere she spoke, calling her a misanthropist, or a charlatan in the local papers.[15] Indeed some colleagues in the Biology Department at my home institution were very displeased when she visited because she denounced GMOs. Those colleagues' research is supported by corporations who patent life for the payout, and grow ecological and economic monocultures.

Shiva's detractors labeled her misanthropic because they perceived the only way to feed the burgeoning population of the Earth was to grow more food. Growing food efficiently on a large scale might be economically sound if we measured only the gross tonnage of food, or resultant Gross Domestic Product (GDP). Such an approach, however, externalized the ecological and social costs of high-input high-yield agriculture, so the market did not tell the truth about the impacts of unsustainable agriculture. Even if full cost accounting—which includes ecosystem services and biodiversity—was implemented, the control and distribution of food resources was not, as many would have us believe, justly allocated.

The controversial economist Garrett Hardin made an argument in the late 1960s and early 1970s which illustrated the dangers of a World Food Bank which sent aid to every famine-stricken village that appeared. Imagine, Hardin said, if a population driven to famine could return to the carrying capacity of their habitat naturally, rather than rush aid to them. Of course, this would mean mass suffering and death for many people who might be saved with outside aid. Without that external input of food, said Hardin, the famine would eventually pull the population in any given ecosystem back into equilibrium (Hardin 1968, 1974). All the Food Bank did, said Hardin, was allow an already poor population to keep reproducing where and when resources were scarce (see Chapter 7, also). Hardin's solution was a sort of green authoritarianism. Each nation, Hardin believed, had finite resources, and should treat its resource base as a metaphorical lifeboat (Hardin 1974). Decisions regarding the distribution of these resources on the life raft would require a lead decision-maker, a "captain" of the boat who made decisions based on lifeboat conditions. To prevent what he called the "tragedy of the commons," he argued for privatization of those resources to prevent their overexploitation.

Hardin had important non-empirical assumptions which drove both his research and his writing. First, he embraced the capitalist notion that decisions were made by rational individuals in a sort of cost-benefit analysis. If those individuals were bounded, discreet, cost-benefit calculators then it would obviously follow that any common resource would be quickly used up, as everyone attempted to maximize their own individual utility, while sharing the cost with the other individuals who held it in common (Hardin 1968). Hardin's fundamental assumptions about the nature of the human and human societies worked in a classical capitalist model, but when it came to real-world applications, they did not hold up to scrutiny. Elinor Ostrum's extensive research compared decision-making regarding common resources in traditional and Indigenous populations (Ostrum 1994). To distill out the portion of her research that is relevant here, Ostrum essentially demonstrated that humans-in-community were *not* bounded, discreet, cost-benefit calculators. They did *not* always make decisions for their own personal well-being.

Especially when sacred values were at stake, people did not make completely "rational" decisions (Atran & Medin 2008; Atran 2011, 2014; Medin & Bang 2014; McCoy et al 2017). Moreover, Hardin did not at all address issues of justice except to say that American Indians, for example, and other marginalized people did indeed have their lands stolen, were excluded from participating politically, and that there was nothing much that could be done about it anyway. Making privatization the solution was precisely the sort of logic that drove Western corporate agriculture, the target of Shiva's critiques. Shiva argued that "The sacrifice of people's rights to create new property rights is not new. It has been part of the hidden history of the rise of capitalism and its technological structures. The laws of private property which arose during the 15th and 16th centuries simultaneously eroded people's common rights to the use of forests and pastures while creating the social conditions for capital accumulation through industriali[z]ation" (Miles & Shiva 1993: 242). Those left out were the poor who, after generations of growing and saving their own seeds suited to local habitats, had to buy them from corporations. To make it worse, those farmers then had to purchase the fertilizers and pesticides to make crops grow in lonely, monocropped fields. Hundreds of thousands of farmers could never work themselves out of debt, and there were hundreds of thousands of suicides among Indian farmers who were caught in that vicious cycle, trying to get out of debt, hoping for a miracle growing season which never came. Bio-communitarianism, not authoritarianism, was Shiva's recipe. An Earth Democracy. As this movement spread it manifested in civil society networks of grassroots resistance to national governments and foreign agricultural monopolies.

Shiva traveled widely to spread her message for two primary purposes. First, to expose the imposition of high-input industrialized agriculture and its effect on the already marginalized people in India. But her struggles resonated with people in similar exploitative economic arrangements across the globe. Shiva brought the social impacts of industrial agriculture into global awareness. Second, she criticized the agents of corporate globalization by highlighting the ecological consequences of monocrop agriculture and its accompanying retinue of fertilizers and pesticides. Diverse agricultural practices promoted ecosystem resilience in an already deteriorating environment.

Like other women who popularized ecofeminism and mounted serious criticisms of economically driven policy solutions (see also Chapters 4, 6, and 7), Shiva specifically targeted the *social* dissolution that resulted from capitalism. The preservation of traditional seeds and planting strategies was a direct affront to the conductors of corporate globalization. That did not stop the spread of Shiva's message, and the legacy of Chipko and Navdanya was the dissemination of grassroots education and assistance with food production. Shiva's activism mostly promoted the spread of a light green environmental religion concerned primarily with the poor, using tactics that re-engaged with the land to use food as the medium for justice. It was an environmentally engaged religion focused on the connection between the life contained in seeds and the well-being of humans. Deep emotionally grounded bonds with the landscapes returned, which had withered in the sterile ecosystems of Western food production. At this writing, Shiva's advocacy continues at Navdanya farm, and the muffled voices she sought to amplify will continue to undercut "rational" and "efficient" food production and grow greener fields for the next generation.

QUESTIONS FOR DISCUSSION:

- How would you explain to a peer what Shiva means when she discusses a monoculture of the mind?
- The Green Revolution increased crop yields in the short term but created long-term impacts on both the soil and the farmers who tilled it. Is exporting agricultural technology a positive or negative practice? What criteria would you use to decide?
- Do you think an Earth Democracy, in the way Shiva advances, is globally possible? Why or why not?

Notes

1 See https://www.navdanya.org/component/content/article?id=616, accessed 5 June 2023.
2 For a critical perspective see D. Harvey (2011).
3 The companies were Monsanto, Syngenta, Dupont, Bayer, BASF, and Dow Chemical.
4 For instance, for the theologian Rudolph Otto the common element was an encounter with the *mysterium tremendum*. For Heinrich Schleiermacher the essence was the feeling of absolute dependence. Gerardus van der Leeuw called this essence a "power," while Paul Tillich described it as the thing of ultimate concern and ground of being. The psychologist and philosopher William James referred to a common subjective "religious experience." Likewise, Mircea Eliade's experience of the non-homogeneity of space was the essence of his thought (see also Chapter 3).
5 Here McCutcheon was criticizing essentialist approaches and the "world religions" approach to religious education.
6 The "Axial Age" was a moniker given to the period beginning in approximately 800 BCE when most of what are now the "world religions" emerged. The prophets and founding figures of Buddhism, Jainism, Judaism, Confucianism, Daoism, Christianity, and Islam were imagined as the key figures of that age. The term was first used by the philosopher of science Karl Jaspers. It is a problematic idea to the extent that it celebrates those religions as *the* religions which were destined for humanity. It was a term that gained wide traction in scholarly circles, however.

7 The reasons for this transition are somewhat hazy still, but it is a question that archaeologists are getting closer to solving every day. What is clear is that it was *not* because of some special cognitive capacity that humans developed, or even more suspect is the notion that humankind was somehow destined to be an agriculturist. Provocative findings from paleochemical analyses indicate that the reason for domestication might have been religious. Intentionally sedentary agricultures might have been necessary for the continuation of religious rituals led by ritual specialists who were herbologists, ancient potion makers. The potion they made was beer. In fact, the so-called agricultural revolution should more appropriately be called the "beer revolution," since it was the use of grain to make potions that precipitated the spread of sedentary subsistence. Some scholars have suggested that these original grain-based brews were entheogenic, which means that they were intentionally psychoactive, even hallucinogenic, since the ergot fungus, endemic to grain crops, is a powerful hallucinogenic compound (Muraresku 2020).

8 In 1991, Iraq invaded the tiny nation of Kuwait in the Persian Gulf. The United States subsequently invaded Iraq. If Iraq had been allowed to annex Kuwait, Iraq's dictator Saddam Hussein would have controlled a significant portion of the global oil supply. This, said advocates of the invasion, compromised national security. In his State of the Union address on 29 January 1991 George Bush, Sr. invoked the image of a "new world order" in which the rule of law would prevail, and dictatorial regimes would be scattered in the face of democratic change. Of course, this was in the wake of the collapse of the Berlin Wall, and the Soviet Union, which left a world where the only remaining superpower was the United States. See https://www.c-span.org/video/?525425-10/a-world-order, accessed 8 June 2023. It was perceived by many as a cloaked vision of American dominance.

9 The five previous mass extinctions occurred 440 million years ago, 365 million years ago, 250 million years ago, 210 million years ago, and 65 million years ago.

10 See https://www.americanbar.org/groups/crsj/publications/human_rights_magazine_home/we-the-people/we-the-people-corporations/#:~:text=One%20of%20the%20strangest%20twists,Clause%20of%20the%20Fourteenth%20Amendment, accessed 8 June 2023.

11 The scholar of religions George James (2013) published a book titled *Ecology is Permanent Economy: The Activism and Environmental Philosophy of Sunderlal Bahuguna*, which explores some of these issues in greater depth.

12 See https://navdanyainternational.org/wp-content/uploads/2022/04/Earth_Democracy_Connecting_RM_HR_WB.pdf, accessed 1 September 2023.

13 See https://www.navdanya.org/earth-university/connecting-rights-of-mother-earth, accessed 11 August 2023.

14 See https://www.navdanya.org/earth-university/connecting-rights-of-mother-earth, accessed 9 June 2023.

15 We experienced this barrage of negative press pieces when she visited my home institution.

10

JANE GOODALL

"Morality Untamed" (Biologist, UN Ambassador of Peace)

In 2006, I was the firekeeper at a *hanbleceya*—an extended Lakota ceremony focused on "crying for a vision." The ritual began with the ceremonial fire. Porous rocks, referred to as grandfathers, were placed in the fire for hours. An *inipi* ceremony followed, where participants entered a round and completely darkened structure with the grandfathers where they sweated to purify themselves for the coming ordeal. The mendicant then retreated to the place where they fasted and sought for a vision, usually upon a hill or elevated terrain. Through four days of fasting with no food and no water, the goal was that the individual received a vision. At its completion, another sweat lodge experience wrapped up the ceremony. The mendicant then communicated their vision to an elder for interpretation. Importantly, the original ceremonial fire must burn through the consummating sweat lodge ceremony—roughly four days. I was admonished that should the fire extinguish, I would be placing the person on the hill in uncertain spiritual territory. I put my sleeping bag on the concrete slab in a pavilion next to the fire, on a point that overlooked a small valley, with the hill on the other side. On the first night, an unexpected and very dramatic lightning storm crashed in, winds whistling around the pavilion, complemented by a drenching rain. The storm lasted two and a half days, but I kept the fire going, with a little help.

My aid came in the grotesque form of a homeless cat who was familiar to the locals, and was aptly called *Igmu*—the Lakota word for "cat." We met during the first night's storm. I woke to find the smoky-colored, one-eyed feline face near mine. Igmu was homely, aloof, always resting just beyond arm's reach. Even so, here was another creature with whom I shared a set of experiences that, at least in my mind, bonded us. He gave me someone to talk to as the storms kept me awake all night. At the end, when it was

DOI: 10.4324/9781003364801-11

time for me to depart, I said a litany of nice things to the cat, who remained stoic and indifferent to my presence.

Encounters with other-than-human persons, other animals, appeared cross-culturally at the center of environmental social movements worldwide, and Jane Goodall is one of the primary vectors by which many people were exposed to the importance of wild nature. Goodall's story is remarkable. The world-famous anthropologist Louis Leakey hired Goodall for her first job, perhaps sensing her brilliance. She founded the Jane Goodall Institute, she was named a United Nations Messenger of Peace, and received a quiver-full of other well-deserved awards for her tireless efforts. Goodall is perhaps the most important voice for animal welfare and conservation in the late twentieth and early twenty-first centuries. For scholars who find their work at the nexus of environmental religions *in* culture, understanding her work is important for several reasons:

1 Her discovery and the dissemination of the finding that chimpanzees not only *use* tools but *make* them as well decentered humans who were imagined as *Homo faber*—the only species that used tools.[1]
2 Goodall and some of her collaborators made the first sophisticated and scientifically documented studies of the emotional lives of other-than-human animals and thus had an impact on scientific methodology, animal testing, and eating animals.
3 She claimed that she recognized animal proto-spirituality in her early work, but in the twenty-first century, she was increasingly vocal about animals having their own spiritual experiences. They help us understand what it means to be human.

Scientific Reasoning Need Not Require Anonymity

Goodall was, by 1960, in the field studying chimpanzees at Gombe Park, but she did not proceed in the manner customary in her profession. She sometimes slept in the forest and walked among the chimps, and while not an insider to the kin group, she was gradually accepted as a nosy neighbor. One of the common practices she disregarded was the typical scientific procedure of assigning animal subjects numbers. How, Goodall protested, could you study such socially attuned primates by labeling them with numbers? Naming was an important component of both Indigenous wisdom and history, since places were often named for events that happened there that impacted the people. People named places that mattered to them.

Meanwhile near Mt. Kenya, Goodall gave the chimps she studied names, she learned their habits, their roles in the local chimpanzee culture, and over the years witnessed their fates. In this instance, Goodall ceded some of *her* power by treating the chimpanzees as *individuals,* something new to the scientific community, which often imagined animals as insensate and expendable. Within ethology, the study of animal behavior, her convention of giving names to the entities she studied, rather than numbers, was eventually adopted by many scientists, and as Goodall reflected, was not only widespread but the typical practice by the turn of the millennium (Goodall and Bekoff 2014).

The reactionary attacks on Goodall were similar to the disinformation and character assassination directed at the biologist Rachel Carson, whose world-famous book *Silent Spring* (1962) blew the whistle on the dangers of chemical pesticides and fertilizers. One of many side effects of widespread chemical use, Carson discovered, was the thinning of birds' eggs. The big agricultural corporations made a concerted effort to discredit her science and characterized her as overly emotional and sentimental. Precisely the sort of character defamation that Goodall weathered. Goodall, then, self-consciously challenged the norms of her own discipline. Fighting then against both critics in her own field and against widely held perceptions of human exceptionalism, she advanced her own global-scale spirituality of hope. As she noted, "When I began my study at Gombe in 1960 it was not permissible—at least in ethological circles—to talk about an animal's mind. Only humans had minds. Nor was it quite proper to talk about animal personality" (Goodall 1990: 14).

Naming her research subjects was just the first move in the long game to reshape her academic discipline. In her book *The Shadow of Man* (1971) Goodall described some of the observations which demonstrated that chimpanzees were both tool-*users* and tool-*makers*. For example, Goodall said: "They use leaves to sop up water they cannot reach with their lips—and first they chew on the leaves and thus increase their absorbency. One individual used a similar sponge to clean out the last smears of brain from the inside of a baboon skull" (Goodall 1971: 240). At the time of her writing, many animals were thought to feel no pain (indeed, in the Middle Ages people thought the same about human infants and toddlers). Because she spent most of her time in the field, her understanding of chimpanzee societies was nuanced. Much of their behavior, she noted, was human-like. When two members of a group encountered each other, "they may bow or crouch to the ground, hold hands, kiss, embrace, touch, or pat each other on almost any part of the body, especially the head and face and genitals." Humans, she noted, demonstrate the survival of such greetings into the *Homo* lineage: "in many cultures, [humans] show one or more of these gestures. Even the touching or holding of another's genitals is a greeting in some societies; in fact, it is described in the Bible, only it has been translated as the placing of the hand under the companion's thigh" (246). Each of those chimp behaviors also transmitted social information. An individual's response was always, whether an attitude of deference or defiance, appropriate to their rank in the social structure. Goodall was among the first who openly talked about the minds of animals. Moreover, she demonstrated that beyond making their own tools, different chimpanzee troops had different technological adaptations that were specific to that group, and which had been passed down culturally to offspring (Goodall 1990: 19).

The chimps, however, faced danger from many directions. Developers, often in league with bureaucratic officials, wished to shrink the reserves in which the chimps lived, or to fragment them, denying the chimp's access to some areas. There was a large consumer demand for bushmeat—flesh from the great apes—so locals had an incentive to disobey legal protections for a profit. Unfortunately, she concluded in 1990 that "It has been estimated by those most familiar with the trade that between ten and twenty chimpanzees will die for every infant that survives to the end of the first year at its ultimate destruction" (Goodall 1990: 221). Just 20 years prior, Goodall made the same point, but her estimation was that six chimpanzees died for each infant stolen from its mother. The trend was clear.

Chimpanzee Religion

I almost called the title of this section, "Chimpanzee Religion?," with the question mark meant to illustrate the tentative nature of the concept of chimpanzee religion. The notion that there were other plant and animal *persons* gained broad acceptance among those in the dark and light green environmentalist milieus, and among many scientists who advocated the inclusion of *empathy as data*. Although those tethered to particular institutionalized worldviews had a narrow understanding of religion, construing it as some private, interior, subjective faith or belief, Goodall sought to capture a primal, perennial religion. One that was accessible to anyone who considered their relationships with other humans and other animals to be morally charged. This was another significant move in the long game of righting relationships with our animal kin. She stated that

> In [their] long quest for truth the scientist has never been able to provide a platform for [hu]man's ancient belief in God and the spirit. Yet, who in the silence of the night or alone in the sunrise has not experienced—just once perhaps—a flash of knowledge 'that passeth all understanding'? And for those of us who believe in the immortality of the spirit, how much richer life must be" (1971: 251-252).[2]

Although evolution has not provided us with a neatly packaged explanation for the emergence and persistence of religions among the great apes,

> "when humans come to an exchange of emotional feelings, most people fall back on the old chimpanzee-type of gestural communication, the cheering pat, the embrace of exuberance, the clasp of hands. And when on

these occasions we often use words, we often use them in rather the same way as a chimpanzee utters his calls.

(Goodall 1971: 249)

What she meant was that on an emotional level words were often onomatopoetic; they conveyed moods and emotions rather than information about the world, and in this, we were not so different than our primate kin. In no uncertain terms, Goodall defended the notion that there is some*thing* beyond our ken, something that qualified as *spiritual* even if its emergence was constrained by cultural idiosyncrasies.

Goodall related her observation that her chimpanzee teachers sometimes expressed elevated emotions when encountering things such as streams, or waterfalls. When her troop approached the Kakombe Stream waterfall "Were the chimpanzees expressing feelings of awe such as those which, in early [hu]man[s], surely gave rise to primitive religions, worship of the elements? Worship of the mystery of water, which seems alive; always rushing on, yet never going; always the same, yet ever different?" (Goodall 1990: 242). She was not talking about institutionalized religion. Goodall constructed a spiritual identity that was beyond the exclusivist labels deployed by most religious groups. For her, spirituality was grounded in felt experience, accessible to all great apes, only bound in the human dimension by the channelization of cultures into hierarchical, specialized, and increasingly urban-dwelling populations. Indeed, after losing her longtime companion Derek, Goodall returned to the Gombe for healing, for spiritual renewal. That time, she said "I was with them not to observe, to learn, but simply because I needed their company, undemanding and free of pity. And as my spirit gradually healed, so I became increasingly aware of a new intuitive empathy with the chimpanzees, with these closest living relatives of ours. Ever since I have felt more in tune with the natural world, the endless cycles of nature, the interdependence of all living things in the forest" (Goodall 1990: 242).

An illustration of the growing credibility of a worldview grounded in compassion, empathy, and belongingness was her inclusion in the Millennium Peace Summit. She recalled that "I was invited to participate in the…[s]ummit of religious and spiritual leaders at the United Nations—over a thousand delegates from one hundred countries representing almost all the major religious and spiritual traditions. I was there to provide a voice for the 'animal nations'" (Bekoff & Goodall & Bekoff 2005: xvii). As others have illustrated, emerging green worldviews are increasingly visible and efficacious. Bron Taylor tracked the roots and expressions of what he termed *dark* green religions (Taylor 2010). They are visible in international political negotiations, films and media, and other civil society-based religious cultural productions (Johnston & Taylor 2016; Taylor 2013; Taylor & Johnston 2016). Despite a number of high-profile statements and documents, there is a growing pile of evidence that environmental behavioral change has not occurred at a large scale within the so-called world religions. However, religious manifestations on the margins of the world religions, and outside of them, have been the most productive areas in which religion and environmental concern intersect (Taylor et al. 2016; Taylor 2010; etc.). For Goodall, acknowledging the deep evolutionary propensity for seeking a sense of purpose greater than oneself provided perspective: "the study at Gombe has done far more than provide material upon which to base our speculations of prehistoric human life. The opening of this window onto the way of life of our closest living relatives gives us a better understanding not only of the chimpanzee place in nature, but also about [hu]man's place in nature" (Goodall 1990: 207).

Interestingly, Goodall and her collaborator, ethologist Mark Bekoff, related their discussion of primate spirituality to a scientific paper by Shunzo Kawamura on the behavior of Japanese macaques on Koshima Island. Kawamura described a female macaque who discovered that the sweet potatoes she ate tasted even better if they were washed in the sea, removing the sand. The other members of the group quickly picked up on her example, and before long all of them were enjoying salty, rather than sandy, sweet potatoes. This story was first published in English in Lyall Watson's *Lifetide: A Biology of Consciousness* (1979). The self-help author Ken Keyes learned of the story through encounters with New Age and Humanistic Psychology advocates. While the veracity of the story is in doubt its message resonated loudly within the anti-nuclear and environmentalist social movements by the 1960s. The Hundredth Monkey story was intuitively appealing whether or not it was factual. Bron Taylor characterized it this way:

> The *Hundredth Monkey* transmits a nature spirituality expressing metaphysics of inter-connection, namely, the belief that, at least at a sub-atomic level, everything is connected to everything else. At the deepest levels of Being, therefore, there is no 'us versus them,' a point which Keyes made explicitly in his book *The Hundredth Monkey*. The metaphysical inter-connection of all universal energies makes possible diverse modes of communication not observable scientifically, including telepathic pathways. They even make possible dramatic and rapid human cultural evolution that could bring social justice, peace, and environmental well-being.
>
> *(Taylor 2005)*

The phenomenon intersected with some New Age and Transcendental Meditation movements, where it was expected that a critical mass was reached, a point at which the behavior of *one more monkey* (or one more transcendental meditator) produced a positive effect through nonempirical telepathic connections. Of course, the macaques, somehow closer to nature than humans according to some, were more readily able to connect to this source. Taylor noted that his "interviews and conversations suggest that the plausibility of the Hundredth Monkey story resides not in its accuracy or scientific credibility. Such stories are resilient within the environmental countercultures because they cohere with the personal spiritual experiences of connection and extraordinary communication that many of these activists have had with nature's various energies and life forms" (Taylor 2005).

None of these expressions of spirituality have much to do with the so-called world religions. Rather they demonstrate that many ideas which circulate in countercultural milieus cross-pollinate and sometimes are received by well-positioned people who might deploy them as motivating metaphors or tropes (Campbell 1972; Taylor 2002). Thinking about religion and the environment requires more than speculation about how religions respond to environmental crises. It also requires the sociological analysis of how these influential ideas are adopted, and adapted in different contexts, and the inclusion of movements that are not all about religion.

Empathetic Encounter and Negotiation

In *The Ten Trusts* (2002), Goodall and Bekoff asked readers to imagine what it might be like to *be* another animal. With sensory capacities very different from humans', would they experience the world much differently? What might we see through the eyes of a bat, or a cat, bear or bonobo, or bee? Some activists have asked what it might be like to *perform*—to be in the world—from the perspective of another animal. Readers may recall Graham Harvey's animistic mantra (see Chapter 1) that we live in a world populated with persons, only some of whom are human, a sort of environmental etiquette (Berkes 2003; Harvey 2006; Nelson 1983; Weston & Cheney 1999). If the world was perceived as being comprised of persons, then affectively rich interpersonal encounters with other-than-human animals may be formative for our own practices and behaviors, which in turn shape our ideas.

The always provocative biologist Richard Dawkins provided one of the most controversial explanations for the emergence of human morality in his now-famous book *The Selfish Gene* (1976). Dawkins suggested that cultural learning units, which he called "memes," could themselves be objects of selective pressures.[3] Memes, however, were selfish, their overall purpose was blindly perpetuating the individual's genetic sequence in the gene pool. Thus, selfish genes supposedly cashed out into selfish memes, which ultimately led to selfish behaviors. The biologist E.O. Wilson's earlier work on *sociobiology* portrayed human morality, religions, and so forth as a product of natural selection. The phiosohper Mary Midgley (1985) and other ctitical voices pointed out some of the philosophical problems with reading human intentionality and affectively associated behaviors (like selfishness) into non-emotive "agents" like genes.

Reactions to sociobiological theory were swift, with noted public intellectuals like Stephen Jay Gould and Richard Lewontin, both biologists, founding the Sociobiology Study Group (SSG) at Harvard University.[4] Their aim was to debunk sociobiological research programs (see the interesting discussion in the introduction of Wilson 2000 [1975]). Those opponents demanded that the human mind was imprinted by the surrounding culture, not entirely determined by biology. Critics of sociobiology worried about the re-emergence of the worst sort of Social Darwinism, where racial or ethnic inequalities might be chalked up to genetic causes, justifying oppressive social structures. Some twenty-first century work addressed those concerns with great care (Kellert 2005, 2007; Kellert & Farnham 2002 Kellert & Wilson 1993)), although sociobiology continues to be a popular target of non-reductionist theorists.[5]

Wilson extended his sociobiological research program by advancing the biophilia hypothesis, which explains why humans have an innate desire for (philia) natural landscapes, sounds, and other living things (bio). Human attraction to other animals and their preference for natural landscapes may be hardwired in the architecture of our brains (Wilson 1984). There are of course bio*phobic* responses to natural stimuli—for instance, human infants typically demonstrate fear of snakes at a very young age. Both tendencies may be evolutionarily adaptive. Biophilia acknowledged the biological basis of human affinities for nature, but also highlighted ways in which social arrangements might recognize and capitalize on those proclivities (Kellert & Wilson 1993). For instance, the sight and/or sound of running water is generally pleasing, and in some cases has been shown to increase productivity when included in a workspace. In some ways, these extensions of biophilia addressed Gould's and Letwontin's desire to preserve a place for culture in the actualization of human phenomes. Biophilia generated research and some strong conclusions that found traction in fields as diverse as philosophy, psychology, theology, landscape planning, and architecture (Kellert & Farnham 2002; Kellert & Wilson 1993; Wilson 1984). Kellert's edited volume with Farnham even included a chapter penned by scholars of religions Mary Evelyn Tucker and John Grim (2002).

Kellert also participated in religion and nature-oriented conclaves, expanding the range of disciplines exposed to and conversant with biophilia. He was a member of the founding Board of Advisors for the International Society for the Study of Religion, Nature, and Culture (ISSRNC) and served in that capacity until his death. His first contributions specifically dedicated to religion and environment came in his contributions to the *Encyclopedia of Religion and Nature* (Kellert 2005), and the first volume of the *Journal for the Study of Religion, Nature, and Culture* (JSRNC).[6] Biophilic encounters have been formative for the moral imagination of the environmentalist milieu. Some of the most well-known examples in the literature are episodes where humans have had highly emotional encounters with other-than-human animals.

Bekoff described his work on coyote populations in Jackson, Wyoming where he was surprised by the precociousness of his subjects. One matriarch called Sally had a litter (for the second year in a row), and Bekoff's aim was to determine how many pups there were, and to observe their process of early socialization. "I found the den," he reported, "and as I approached it, I felt the presence of another animal. I had not seen, heard, or smelled Sally or any other coyote, but I had this unnerving feeling that someone was there. Then I heard a soft bark. I turned around and saw Sally watching me from about ten meters away" (Goodall & Bekoff 2002: 43). The description of his experience recalled for me the episode I had with a bobcat when I was young, which was recounted in the Introduction. People protect what they love, and they love things because of intimate, empathetic encounters. If we do not experience the world *with* other animal persons, can it be protected? Bekoff opined that "We are beginning to learn that each animal has a life and a place and role in this world. If we place compassion and care in the middle of all our dealings with the animal world and honor and respect their lives, our attitudes will change. All lives become sacred and we become more careful with our care of them" (Goodall & Bekoff 2005: 43). Intimate encounters, whether terrifying, awesome, or sublime, are often recounted in environmentalist subcultures, but to the extent that they are highly emotive encounters they are also risky, for in such negotiations with others an individual may be required to compromise on some of their deeply held presuppositions or perceptions. Bekoff recalled a particularly emotional eye-to-eye encounter with an animal research subject called Speedo (a cat) during his graduate work. After tracking performance on wayfinding tasks Bekoff was supposed to surgically remove much of the part of the brain that processes visual sensory inputs, and then track them again. After they had collected the needed data, the subjects were euthanized. Bekoff locked eyes with Speedo during the procedure. It was this event that set Bekoff's career in another direction, finding a place for scholastic pursuits that did not demand such personal compromises in the name of science.

Examples from environmental literature are ubiquitous. The prophet of preservation John Muir (1838–1914) recounted his adventure with a canine companion called "Stickeen." Muir recalled one journey where he ended up in a camp where there lived a dog. The dog often trekked with Muir. One day, as Muir beheld a storm rising over the glaciers, he rushed out to meet it: "for many of Nature's finest lessons are to be found in her storms, and if careful to keep in right relations with them, we may go safely abroad with them" (Muir 1997: 558). Muir and his canine companion crossed glaciers, leaping crevasses when they encountered them, keeping their faces to the storm. Eventually, however, on the way back to camp, Muir realized that they had come to a place completely encircled by crevasses—a glacial island. They made a daring and dangerous escape, with Muir saying that he had never for so long been so close to death. Stickeen's carefully considered bravery was awe-inspiring to Muir, and because of that travail, they were bonded. After that harrowing experience, Stickeen "always lay by my side, tried to keep me constantly in sight, and would hardly accept a morsel of food, however tempting, from any hand but mine. At night, when all was quiet about the camp-fire, he would come to me and rest his head on my knee with a look of devotion as if I were his god" (Muir 1997: 566). He was, Muir opined, "a true child of the wilderness, holding the even tenor of his hidden life with the silence and serenity of Nature. His strength of character lay in his eyes. They looked as old as the hills, and as young, and as wild: I never tired at looking into them" (Muir 1997: 557). There are different endings to Stickeen's story depending on the source, but John Muir, the prophet of wilderness (Nash 1967), had his own eye to eye encounter.

Other tales of similar close encounters followed celebrated naturalists. Readers may recall from Chapter 8 that J. Baird Callicott's inspiration, the ecologist Aldo Leopld, shot a mother wolf and gazed into her eyes as she drew her last breath. "There was something new to me in those eyes," Leopold said, "something known only to her and the mountain" (Leopold 1949: 138). This narrative was broadcast throughout environmentalist countercultures. For example, Dave Foreman, perhaps the most well-known founder of the radical environmental group Earth First!, deployed it as an emotionally rich climax of his traveling stump speeches.

Paul Watson was a Greenpeace activist who broke from the group because they were not willing to engage in more confrontational forms of resistance to animal exploitation. He created the Sea Shepard Conservation Society to harry whaling vessels and shut down both illegal and legal whaling around the globe. He was a controversial figure because he rammed whaling vessels, maneuvered between the whalers and the whales, and eluded capture by staying on the sea, never making port. In one episode, in 1975, Watson and a co-conspirator repeatedly maneuvered their small boat between a pod of sperm whales and a Soviet whaling vessel. Watson's popular speech, repeated in many venues, featured the claim that he had been studying Gandhi at the time. After harrying the whaling vessel, the Soviet captain came out to speak to the harpooner, looked at Watson, and aggressively drew his hand across his throat to signal that Watson was now one of the hunted also. In his speeches, Watson quipped that was the moment he realized that Gandhi's approach would not work for him that day. The next harpoon went directly over Watson's head and plunged into the back of a female sperm whale. Watson continued: "The dominant male surfaced and threw himself at the harpooner. But the harpooner was ready and fired an unattached harpoon into the head of the male at point blank range." Watson recalled the giant whale fell forward that he would fall directly on to Watson's little boat. The whale appeared to summon his last bit of strength to avoid the craft as he collapsed. Writhing on the surface, before he sunk, Watson said he caught the whale's eye. "What I saw in that eye," said Watson, was understanding:

> Understanding that we were trying to help them, they knew what we were trying to do. I could see the effort he made to pull himself back. I saw that eye sink beneath the surface, and he died. So it could have killed us, and chose not to do so. So I'm personally indebted to him for saving my life. But I also saw something else in that eye. Pity, and not for himself but for us. That we could take life so thoughtlessly, so remorselessly. And for what? What were the Russians killing whales for? It wasn't for meat. It was for spermaceti oil.

That oil, Watson continued, was used for high heat machinery applications. In this case, the oil was being harvested for the construction of intercontinental ballistic missiles (ICBMs). Humans were sacrificing this sentient and awe-inspiring creature for the manufacture of weapons "meant for the mass extermination of people. And that's when it hit me: we're insane," Watson said.[7]

There are two ways to imagine, from the perspective of natural selection, why animals behaved altruistically. The first was kin selection, which hypothesized that animals help others who are kin, those with whom we share at least some of our genes. So self-sacrifice for the good of the "selfish genetic sequence" was still beneficial to the animal's lineage. Another explanation is reciprocal altruism, where even in the absence of shared genes, animals helped others because there is an expectation that their act of selflessness will be rewarded, or the kindness returned (Goodall & Bekoff 2005: 56). But it is unclear whether these two perspectives exhaust the breadth of the ways in which animals interrelate, whether *all* instances of animal kindness are explained by those theories.

Empathy and Spirituality

Goodall's own experiences led her to unequivocally declare in her later works that primates did possess the capacity for spirituality and that there are many situations in which it was evident that something deeper than transactional kindness must be behind affective encounters with other creatures. Although Goodall's early works were explicit about the emotional lives of animals (Bekoff & Goodall 2005), it was not until later that she wrote about her own personal spiritual sensibilities. Goodall raised the question, "If the chimpanzees had a spoken language, if they could discuss these feelings among themselves, might not they lead to an animistic, pagan worship of the elements?"

(Goodall 2005). She did not have a final answer to this question, but spontaneous and demonstrative ritual elements like the "waterfall dances," were "precursors of religious ritual" (Goodall 2005).

Goodall recalled that sometimes in her field excursions:

> there were moments of perception that seemed almost mystical so that I became ever more attuned to the great Spiritual Power that I felt around me – the Power that is worshipped as God, Allah, Tao, Brahma, the Great Spirit, the Creator, and so on. I came to believe that all living things possess a spark of that Spiritual Power. We humans, with our uniquely sophisticated minds and our spoken language that enables us to share and discuss ideas, call that spark, in ourselves, a 'soul.' Is not the same true for a chimpanzee? Or any other sentient, sapient being? It is most unlikely, however, that any animals other than ourselves care – or are capable of caring – as to whether or not they possess immortal souls!
>
> *(Goodall 2005)*

In an argument parallel to the famous historian Lynn White (1967) she said that "In the original Hebrew text of Genesis chapter 1 verse 26 God gave man '*v'yirdu*' over his creation, and this has been translated as 'dominion.' But many Hebrew scholars believe that the true meaning of the word is to 'rule over,' as a wise king rules his subjects, 'with care and respect.' A sense of responsibility and enlightened stewardship is implied. St. Francis understood. But throughout the Judeo-Christian world today animals are typically regarded as mere *things*, to do with as we will so long as it is for, or might be for, human good" (Goodall 2005).

Goodall's spiritual experiences and vision are exposed throughout her work, but her book *Reason for Hope: A Spiritual Journey* (1999) performed an extended analysis of her own spiritual perceptions. One day in the field Goodall and her chimpanzee family were out when a thunderous storm overtook them. She huddled in the forest with her family and reported that she "lost all track of time" (Goodall 1999: 172). Reflecting later on that experience, she compared it to the insights of great masters from all corners of the world, "the mystics and the holy men of the East, and the founders of the great world religions" all pondered human purpose (175). "That afternoon," she mused, "it had been as though an unseen hand had drawn back a curtain and, for the briefest moment, I had seen through such a window. In a flash of 'outsight' I had known timelessness and quiet ecstasy, sensed a truth of which mainstream science is merely a small fraction. And I knew that the revelation would be with me for the rest of my life, imperfectly remembered within. A source of strength..." (175). She referred to this as her Damascus experience (264), comparing it to a biblical story in which a Roman tax collector named Saul encountered a vision of Jesus on the road to

Damascus, changed his name to Paul, and spent the rest of his days spreading Christianity across the Mediterranean. Indeed, Goodall was a disciple of the Law of Life which was written in the fabric of the biological community, and she was a tireless evangelist for reminding humans that they are just one species—not a particularly fast, or well armored, nor strong species—among many others, and that we ought to foster acceptance of this fact. All religions, the scriptures, books, and mystic writings came to a common understanding about the world: that humans were born from the world, are a part of it, and not exempt from life's laws.

Others like Frans de Waal (1996, 2009, 2013, 2020) likewise contributed to the increasingly obvious facts: (a) that social groups can have morality without religions, and (b) given animals' modes of empathetic and emotive relations within and between their social groups, they exhibit spirituality sensibilities. Both Goodall and de Waal have participated in meetings which included prominent religious studies scholars and have enthusiastically embraced a budding interest in animals' role in religions. Goodall and her colleague Mark Bekoff (Goodall & Bekoff 2014) highlighted the profoundly empathetic and ethical dimensions of animal behaviors, concluding that their emotional and possibly spiritual lives are much more complicated than we tend to give them credit for.

By 1996, the Religion and Animals Consultation was launched as one of the subgroups of the American Academy of Religion (AAR).[8] The "buzzing, breathing, clicking, clacking, biting, stinging, chirping, howling landscape of religious studies" (van Horn 2011: 230) diversified in the subsequent decades as explorations from the biblical to the behavioral, from the aesthetic to the cognitive have blossomed. As of this writing, there have been a number of instances where particular animals or places have been recognized as having their own inherent worth and right to flourish that are worthy of moral and even legal defense. To what extent will the increasingly frequent deployment of such legal frameworks and their challenges impact environmental ethics, conservation, and capitalism? Stay tuned.

QUESTIONS FOR DISCUSSION:

- Some people appreciate and accept Jane Goodall's depictions of chimpanzee spirituality. Can you think of some of the reasons why other people might reject the notion of spirituality in other-than-human animals?
- Do you find the tale of the Hundredth Monkey provocative? Why or why not? Is collective consciousness-raising possible?
- Have you ever had an "eye to eye" experience with another animal (wild or domesticated)? What animal was it? How did you feel?

Notes

1 The theme that humans are "makers" stretched back into antiquity and was exercised by several different thinkers. It is one of the many (now debunked) ways that humans protected their exceptionalism.

2 This is a reference to the Christian New Testament, Philippians 4:7: "And the peace of God, which passeth all understanding, shall keep your hearts and minds through Christ Jesus."

3 This use of the term "meme" conveyed a cultural behavior transmitted between generations. At this writing, the use of the term meme refers to something different: typically, the creation of, or the mashing together of different media (art, music, and vocalizations) to create a critique or satire of the original sources of the *bricolage*.

4 E.O. Wilson, Gould, and Lewontin all worked at the same institution. Imagine the faculty happy hours.

5 Interestingly, Wilson expressed surprise that while a mere 30 or less pages of *Sociobiology* were dedicated to the implications of the theory for humans, the academic response seemed disproportionately large.

6 Information about the history of the Society and its affiliated journal can be found at https://www.issrnc.org/about/history/, accessed 23 June 2023.

7 This was a frequent component of his speeches. See, for example, https://www.youtube.com/watch?v=YYZnaSr3TII, accessed 6 September 2023.

8 The Animals and Religion Consultation was later expanded to the Animals and Religion Group, which remains active at this writing.

CONCLUSION

"Themes in Religion and the Environment"

The Transformation of Human Relations with Their Habitats

To provide a bird's-eye view of the study of religions and their ecologies, I will review some of the most important tributaries and streams of contemporary debate and discussion. Each chapter illustrates research programs introduced by, or that followed from the featured key thinkers. Here, however, the focus is on four broader themes which emerged in the foundation of the field and its dynamic development over time. (1) First, a theme that runs through most of these chapters is a concern with the mode of production, which appears as a backdrop to many of the methodological and theoretical positions staked out by the key thinkers. That is the subject of this first section. (2) Second, and related to the first, there was a process of ecological and economic disruption, and causes for the perceived disconnect between humans and their environments. New modes of production necessarily reconfigured relationships and as Durkheim noted, a sense of *anomie* emerged when individual well-being was not fulfilled by the new relationships. (3) Third, and related to the first two changes, was the relationship between different ontologies, epistemologies, and normativities. In other words, it is helpful to explore the ways in which individuals and groups gathered empirical data from the world (epistemology), how that knowledge informed perceptions of reality (ontology), and any related moral considerations or obligations (normativity). Certainly, some technological advancements impacted human relationships with the natural world, and many of them predated the emergence of the Western scientific methodology. Western science is simply one mode of gathering empirical data about the world. As it spread across the globe, that one ethnocentric epistemology was imagined

DOI: 10.4324/9781003364801-12

as universally applicable. In reality, however, even when Western scientific perspectives were hegemonic, it was abundantly clear that non-western empiricism played a formative role in the construction of ostensibly Western science (Bauman 2014; Harding 2015, 2002, 1991; Cajete 2016, 2015). It is important for scholars working in this area to recognize that empiricism was foundational to many different historical, social, economic, and ecologic relations, not just Western ones. (4) Fourth, the chapters raise some important questions regarding relationships with the other-than-human sentient creatures with whom we share an ecological habitat. Animality becomes important in conversations about who (not what) we eat. In addition, animals have become an inspiration for new technologies derived from biomimicry. Further, research continues to challenge the various ways in which humans imagine they are exceptional. Humans are not the only tool users, nor the only ones with rituals. We share some of the higher cognitive functions of other apes and monkeys, but also with cetaceans (dolphins and whales) and other creatures, including representatives of the phylum *Mollusca* (octopi and their kin), and corvids (ravens, crows, and magpies), etc. Other animals have emotions, feel pain, engage in social relationships, and exchange energy and materials through complex associational webs. All these shaped and continue to expand the range of scholarship related to religions and their biological and cultural matrices.

The study of religions and environment has long been concerned with the root causes of the divorce between humans and the rest of the biosphere. As this book illustrates, one need not be a religious studies scholar to investigate the ways in which religions, cultures, and their environments intersect. For instance, the human ecologist Paul Shepard and the anthropologist Roy Rappaport were two (and not the only) scholars who traced the "fall" of humanity from its original relationship with nature to the emergence of sedentary agriculture approximately 10–12,000 YBP. The emergence of urbanization was the most significant shift in the mode of production in human societies, spreading quickly from its birthplace in modern-day Turkey, Armenia, Iraq, and Iran. The reasons for this transition are not entirely clear, but several theories were offered to explain the transition.

Some problematic notions included the "intellectual readiness" theory and the "demographic model." The intellectual readiness model posited that humans were suddenly intellectually ripe for the shift to sedentary agriculture. As we now know, however, our genetic blueprint and our brains are largely the product of adaptations which emerged in the Middle Pleistocene era (300,000–30,000 YA). So, there was no sudden onset of cognitive power that was qualitatively different than that which was present before sedentary agriculture. Moreover, such a model assumed that humans have a *natural* proclivity for agriculture. Sedentary agriculture was just one possible trajectory for anatomically modern humans, not the only possible end game. The demographic model posited that sedentary agriculture developed to feed what was at that time a burgeoning human population. Imagining that an expanded human herd would be saved by the development of sedentary agriculture is analogous to jumping out of an airplane and planning to knit a parachute on the way down. Sedentary agriculture is not immediately productive on a large scale, and it requires relatively entrenched social hierarchies and stratification to manage the construction of infrastructure to maintain such a way of life. Increased populations were more likely a result of sedentary agriculture and urbanization, not their cause.

More plausible hypotheses took account of climatic shifts in the Middle Pleistocene (300,000–30,000 YBP) which would have generated new ecotones, areas where two or more different ecosystems intersect—the "marginal habitat model." In essence, such overlapping spaces were often more biodiverse than their respective parent ecosystems because of their more varied habitats. Other hypotheses that may have more explanatory power include the "oasis model," which posited that post-Pleistocene drying following the ice ages exposed some choice areas supportive of more productive agriculture. The "dump heap" model hypothesized that human sedentary agriculture emerged through selecting and cultivating foods that might have cropped up in compost or other food discard piles. Anyone who has kept a static compost pile at home has witnessed the escape of squash, tomato, or other vegetable plants from the compost pile as they decomposed.

It is likely that no single causal element drove the emergence of widespread sedentary agriculture. But prior to this development, human populations acquired subsistence through multiple modes of production, including hunting, foraging, scavenging, and planting. Agriculture was likely practiced in some form for tens of thousands of years before the emergence of city-building and sedentarism. It was not, however, the primary means of food production prior to about 10–12,000 YBP. In the periodization of earth's systems, the Holocene began approximately 12,000 YBP, which included the emergence of Neolithic tool making, the rapidly increasing agricultural productivity, and the spread of sedentarism. The Holocene was a relatively good time for primates who liked warm climates, and human niche construction favored concentrated cultivation from the Persian Gulf to the Black Sea. Sedentary agriculture relatively quickly produced surplus food in such a climate. Basic biological facts dictated that when any population of a given species had an increase in food supply, more population growth followed. Rather than oscillating around a population mean relative to food availability, for the first time the new mode of production allowed humans to bend basic biological laws, increasing food production and population, and annexing more land to transform into sedentary agriculture. As populations and production expanded further from primary water sources, there was an increasing pressure toward the specialization of labor—and that development should not be glossed over. For instance, irrigation was necessary to expand productive capacity to land that was not previously suited to agriculture. Such technological advancements derived from specialization. As populations and food supplies increased, and as more land was annexed, the specialization of labor continued to offer benefits. If one person could now produce enough food to feed not only their own, but their neighbor's family, that neighbor was freed up to work on other problems, like better plows, more efficient irrigation, or the manufacture of weapons. The distribution of a growing food supply and the need to coordinate infrastructure projects drove the emergence of social stratification. Archaeological data illustrated that through the Upper Paleolithic period (30,000-circa 12,000 YBP) human societies appeared to exhibit relatively egalitarian populations—there were no kings, just accountability to kin and extended family. Around 12,000 years ago, however, not very long in geological time, with the rise of hierarchical societies, visible in the archaeological record were larger dwelling places (perhaps palaces) typically connected to a granary, while other dwellings remained of similar size. Thus, agricultural production and specialization become the early agents of urbanization.

"Civilized" Economies and Their Religions

As those societies spread geographically, they inevitably encountered other human groups with different means of production (i.e., shepherds, seminomadic hunter-gatherer groups, or others with mixed modes of production). The societies with sedentary agriculture invariably possessed a larger population due to food surplus, and with the specialization of labor, the possibility of specialized implements for war. There were populations whose warlike advance overtook others with different means of production so their land could be placed under cultivation. The neighboring social groups stood little chance against those overwhelming foes. Ultimately, they had two choices: (a) to move away, which might delay the disruption of their lifeways a while, but not forever, or (b) to adopt the wildly successful sedentary agricultural experiment, thereby gaining the security of allying with other existing agricultural groups. In this way agriculture spread, becoming the norm for increasing numbers of social groups. Our species history is not only tied to, but deeply dependent on the procurement and control of calories from other plants and animals (including other humans). About 12,000YBP, this dependence began to morph into domination of the Earth's ecosystem services and the domination of other humans. Sedentary agriculture was not a "revolution" as it was described in my preparatory education. Rather, it was one particular cultural experiment that, through a series of coincidences, became successful, but at the expense of diversified human adaptations to changing earth systems.

Now here's the interesting part: *All these fascinating developments in human-environment relations were prehistoric.* They all occurred before there were any written records. And perhaps most importantly, what we now refer to as "world religions" were not invented for another 7,000–8,000 years. In other words, the things that many take for granted as "religions" were creations of urban societies where humans lived in sufficient numbers to require the importation of resources. The contexts of all those religions were specific to unique times and places. The point is that human groups lived as sedentary agriculturists and city-builders for thousands of years before anything was written down. At least 200,000 years of human history were counted as pre-history. Only those things that happened *after* agriculture and urbanization were preserved in writing, and in the later emergence of now global religions. There was nobody living at that time who could have had any idea that humans had engineered hundreds, if not thousands, of other cultural experiments with other modes of production before they were overcome by early agriculturalists. All these religions, East and West, addressed *human* problems in a world where ecological limits were breached deep in their prehistoric past. The illusion of human mastery was a foundational assumption of all those religions and the societies from which they grew.

Most of the scholars represented in this volume focused on human relationships with nature and associated religious complexes following the emergence of sedentary agriculture. They focused on civilized religions, ones that emerged after the so-called Axial Age, and many of those religious traditions viewed the present world as fallen. The goal of Dharmic religions (e.g., Hinduism, Buddhism, and Jainism) was to escape this world of suffering, to cease being reborn into it. The goal of some Abrahamic religions (e.g., Christianity and Islam) was an escape from the world in search of eternal life beyond it. In general, all the so-called world religions emerged when demographic pressures, disease, conflict, social stratification, and other forces created significant populations of disenfranchised and suffering people. All of the religions which emerged during and after this Axial Age essentially viewed this world as broken or incomplete, and imagined a coming salvation from it (Quinn 1996).

Rosemary Radford Ruether analyzed the early civilized (mostly urban) religions of the Mediterranean world. Their societies were complex enough to evidence what Durkheim would have termed an organic mode of social solidarity, exhibiting significant specialization. Ruether's scholarship included the evaluation of some of the matriarchal societies and cultures that emerged in the ancient Near East, exposing the increasingly androcentric hierarchy that oppressed both women and destroyed ecosystems. By the time the religions that occupied Ruether's scholarship emerged, the divorce between humans and nature was old enough for people to forget that the environment had been anything other than a passive setting for human procurement of calories. Ruether's work, women's issues in general, and peasant populations and marginalized people specifically, exposed the causal influence of economics in the oppression of women and the Earth. Like Durkehim, Ruether recognized the alienation that accompanied the industrial revolution. To understand the rise of male-dominated societies with organic solidarity, both Durkheim and Ruether commented on the work of the political economist Karl Marx.

Marx's analysis of the alienating effects of capitalism was fundamental to Durkheim's and Ruether's thought. Both thinkers, however, distanced themselves from Marx. For Marx, alienation was the inevitable result of production processes where humans played an increasingly specialized role. In the era of assembly lines, a worker was no longer a craftsperson; they were a cog in the larger productive machine. Alienation was the product of the rigidification of the social system, an abnormal and undesirable condition. Ruether took from Marx the analysis of class struggle. Marx imagined the struggle between the lower class (proletariat) and the aristocratic class (the bourgeoise) was the main driver behind all historic events. The context of Ruether's writing, included peasant and feminist movements which were the result of the alienating effects of capital.

Part of Marx's corpus addressed the shift from feudal to capitalist modes of production and the creation of a peasant class. Feudalism was both a political and social philosophy focused on the relationships between land and labor that ran from the fifth century through the fifteenth century. The aristocracy of the Middle Ages morphed, and feudalism gave way to the Renaissance in the fifteenth through the seventeenth centuries. Modernity emerged out of the Renaissance spurred by the rise of Western scientific reasoning, political movements which supported democratic values, and economic systems which buttressed democratic values individual freedom.

The historian Carolyn Merchant's analyses began in the Renaissance, particularly focused on the impacts of Western empiricism and science. Merchant highlighted philosophical developments including the rise of Western empiricism going back to Francis Bacon and Rene Descartes. Bacon is often cast as the father of the Western scientific method. By his time, however, androcentrism was so enmeshed in the cultural milieu of the colonizing powers that Bacon's contributions easily employed misogynistic metaphors and tropes to describe nature as a recalcitrant female whose secrets must be forcibly taken from her, so that humans might exercise control over the processes of life. Much of Merchant's work focused on the intellectual milieus which precipitated and perpetuated the separation of humans from nature and the denigration of both nature and women. Her specialization on the emergence of Western scientific methodologies meant that she also reflected on the role of technology in the creation of these binary oppositions.

The historian of science and technology Lynn White, Jr. (whose arguments were alluded to in several different chapters) likewise traced the "fall" of humans to the emergence of Western science and technology. For him, more and more invasive agricultural technologies coupled with increasingly domination-oriented theologies were the direct causes of the ecological crisis (White 1967). In both Merchant's and White's narratives, it was impossible to tell which way the causal relationship ran. Did increasingly dominion-oriented theologies foster more invasive food production practices? Or did more intense forms of agriculture create theologies to reflect the growing human occupation of earth systems? We might surmise simply that there was a positive feedback loop between these two—the technological and the theological. But those were the sorts of conundrums that scholars and students of religion and the environment pursued.

Other scholars present in this volume criticized the social disparity that was a necessary product of the sort of capitalism that became dominant during the modern period. The Lakota scholar Vine Deloria levied powerful indictments of capitalism and its impacts on tribal peoples in the North American context. The mutual reinforcement of individual liberty and capitalism spelled danger for Indigenous populations whose religious practice and ideas of the sacred were not easily reconciled with interpretations of religions rendered in courts of law. Life-affirming values were destroyed as capitalism took root in the global South, often in places which still bore the scars of colonialism.

For example, chapter 9 on Vandana Shiva illustrated the deleterious social and biological effects of the so-called Green Revolution. In India, the social effects were obvious when genetically modified seeds replaced locally adapted crops, and farmers were forced to go into debt to procure the new GMO seeds. Hundreds of thousands of suicides because of inescapable debt cycles, and the loss of traditional familial roles, were just some of the forms of social fallout. Biologically, those genetically modified seeds required the almost complete disruption of ecosystems. The diverse crops that were grown by rural farmers were replaced by just a single crop—a monocrop scheme. The simplified ecosystem lacked the inherent benefits of diverse agronomy, like natural pest defenses, or resilience in the face of disturbance. In monocropped fields, a pest species could spell disaster without habitats for natural predators. To alleviate this concern, the genetically modified crops were engineered so they were resistant to powerful pesticides and herbicides. The herbicides killed any plant that was not the cash crop, and the pesticides and fungicides took care of the dangers from animals and fungi. However, because the soils were not replenished seasonally as typical fields were, they required additional high inputs of fertilizer to ensure continuing yields. Wangari Maathai likewise noted the negative impacts of Western-style agriculture on the desirable species upon which women had depended for millennia. As in India, women in Kenya, and later in other African nations, attempted to restore some traditional modes of land use and production. In both cases, the shepherds for this ancient wisdom were women who, despite the corrosive effects of capitalism and Western science, struggled to better the lives of peasant populations. Common to many of the thinkers featured in this book were strong criticisms of the globalization of Western science and global capitalism, which compromised the health and well-being of many populations across the globe, especially those who were already marginalized. In addition, capitalism facilitated the liquidation of resources for profit without regard for the environmental impacts of such development. Both social and ecological consequences were clear. Importantly, the prevailing mode of capitalism is not the only form that it may take. Scientists and economists, sometimes in league with theologians and religious leaders, advocated capitalist forms that might be more inclusive and less ecologically destructive (Daly 1973, 1996; Daly & Cobb 1989; Odum 1970; Odum & Odum 1976, 2001; Sen 1999).

In the late twentieth and early twenty-first century, some evangelical Christians attacked environmentalism as a dangerous, corrosive, pagan religion. Environmentalists were cast as misanthropic—they supposedly disliked if not hated people and society, and environmental efforts would hurt the poor the most (Veldman et al. 2014). Other environmental religions have emerged as existing religious institutions, groups, and people reinterpret their traditions or develop novel practices or liturgical elements in environmentally beneficial ways. Still, other environmental religions are entirely new religious movements that advocate varying degrees of environmental care. In all cases, the emergence of capitalism has driven differential access to the procurement and distribution of resources—it manufactured social disparity. Moreover, it made it more difficult for those who did not share Western religions, ontologies, and epistemologies to survive.

Multiple Ontologies and Epistemologies

The early thinkers reviewed here, E.B. Tylor, Emile Durkheim, and later Mircea Eliade (and their intellectual kin), took for granted the notion that humans were born to become agriculturalists and city-builders. The notion that there were other ways humans had lived in the past, indeed thousands of different cultural experiments, was lost in the darkness of "prehistory." Any cultures surviving from pre-historic times (i.e., Australian Aboriginals, the San of the Kalahari, and others) were imagined as primitive, backward, and savage. that as societies matured, they would outgrow their animism, variously referred to by scholars as totemism, shamanism, or other terms for the primordial religion, and grow toward monotheism. According to Tylor and Frazer, monotheisms, while certainly more rational than animism or polytheism, were destined to be eclipsed by science. Tylor's and Frazer's perspectives were shaped by the impactful spread of Darwin's theory of species evolution, applying *biological* principles to *cultural* evolution. Others, like Durkheim, explicitly resisted explaining social formations in Darwinian terms, worrying that such an interpretation reflected badly on those more ancient cultures. Instead, he argued that they merely expressed a different mode of social solidarity. However, Durkheim's differentiation between mechanical and organic solidarity still intimated that those arrested in mechanical solidarity were not capable of navigating the conditions of organic solidarity. Even if Durkheim wanted to demonstrate that biological evolution was not an accurate portrait of cultural development, he and others of that era assumed that Western science provided the most accurate descriptions of the world. Others in this volume, however, possessed the vision to see that Western ontologies and epistemologies were not universal.

Rosemary Radford Ruether and Carolyn Merchant both highlighted the complicity of Western scientific reasoning in the derogation of women and the despoilation of the environment. Especially Merchant unpacked the male-centered logic of human exceptionalism—the notion that humans were the only (or the primary) creatures with moral standing. Ruether and Merchant put Western scientific logic into a larger historical framework. Ruether developed a theology that honored the matriarchal roots of civilizations in the Near East, placing the development of male-centered societies and theologies into a broader historical perspective. Merchant's work focused on the cultural products of the Renaissance and the Western scientific method, perceiving those developments as destructive and oppressive, targeting women and nature as things that must be domesticated to be useful. Although they traced male-centered oppressions during different historical periods, both placed the emergence and exercise of Western science into a broader history of the erection of patriarchal social organization.

As noted above, both Vandana Shiva and Wangari Maathai battled against the twentieth and early twenty-first century imposition of Western scientific agricultural solutions on rural, traditional populations who had long managed resources through religious complexes that ensured their provision over the long term. In Shiva's analyses, Indian farmers were subjected to corporate control of seed stocks and farming methods. In Kenya, Maathai chafed against male-dominated governmental administrative schemes modeled on Western resource management, arguing that they framed the work in technical terms and focused on the production of resources in global markets. In contrast, Maathai fostered networks of "farmers without degrees" who utilized traditional management concepts and practices. They managed to sustain their livelihoods and to resuscitate dormant, women-led conservation efforts.

Vine Deloria also questioned the supremacy of Western knowledge systems and related concepts such as *religion* and the *sacred*. Basically, Deloria argued, Abrahamic religions traditionally downplayed the experiential aspect of religion, privileging instead intellectual commitments to invisible deities. In one important historic case, *Employment Division v. Smith* (1990), Supreme Court Justice Anton Scalia, a Catholic, proclaimed in his decision that "It would doubtless be unconstitutional, for example, to ban the casting of 'statues that are to be used for worship purposes,' or to prohibit bowing down before a golden calf'" (Deloria 1999: 215). Those practices were obviously protected. The Native peoples' use of a sacred cactus in ceremony was not. It was ironic, Deloria noted, that one of the forms of religion protected by the U.S. Constitution was a pagan ritual directed toward a statue (in the biblical book of Exodus the people craft and worship a golden calf). In the case of *Employment Division v. Smith*, the court ruled that idolatry was protected, while the ingestion of ritual elements that were considered federally prohibited substances, was not. "In reality," Deloria argued, "the white man's myths are no better, or more accurate than ours. They believed we came from the lost tribes, from Noah's ark, or whatever.... any group that frantically dug gold in the west in order to transplant it to the east and bury it cannot be quite right and their insights cannot form the highest achievement of our species" (Deloria 1999: 136). Deloria attacked Western history and science which masqueraded as objective, value free. He reminded readers that the supremacy of Western ideas erected a façade of universalism. But the world did not uniformly adopt those supposedly universal norms. As Deloria put it, "The recognition that there is no homogeneous sense of time shared by all societies must certainly become apparent to us if it is not already clear. We can and must, therefore, create a new understanding of universal planetary history" (Deloria 2003: 64). Native and Western epistemologies and ontologies, their knowledge systems and perceptions of reality, were fundamentally conflictual. Environmental religions and their associated social movements were examples of resistance to typical Western beliefs and superficial values, and the search for authenticity in Indigenous religions (Deloria 1992: 52).

J. Baird Callicott also advanced Indigenous religions as potential correctives to dualistic, human-centered worldviews and values. His work on mining the conceptual sources of existing religions to find and amplify glimmers of green included many examples of Indigenous or traditional religions (Callicott 1994). Callicott along with the philosopher Michael Nelson also examined traditional Ojibway environmental ethics, drawing from their myths, narratives, and practices (Callicott & Nelson 2004). Callicott's and Nelson's treatment of cultural relativism, the notion that different cultures had different ontologies, epistemologies, and axiologies (ethical systems), echoed Deloria's arguments regarding the false universalism of Western perspectives. Callicott's idea of worldview remediation depended on embracing the findings of what he referred to as the New Ecology and the New Physics. The systems perspective required to understand these sciences was also relevant to overcoming the mind/body duality that contributed significantly to environmental crises. Callicott, like Deloria, referred to the concept of quantum entanglement to argue that the world was not comprised of a collection of objects (Deloria & Deloria 2006: 97, 155–160). They envisioned the world as comprised of a communion of subjects. The fundamental organizing feature of the world was relationship—no feature of an ecological system could be abstracted out; it could only be understood in its totality. The recognition of this relationality was present in both Deloria and Callicott's challenges to Western ontologies and epistemologies.

Each of the thinkers above argued that the preservation of traditional ecological knowledge (TEK), ancient perceptions of the relationships between humans and their habitats, was crucial to addressing ecological crises. Scholars in the Anglophone world began to attend to Indigenous and traditional taxonomies and natural categories, all of which were interpreted by Western scholars as entangled with religions. Ethnobotany, the study of botanical (plant) species and taxonomies outside the Western scientific knowledge, was a new field of study by the late 1800s. Well-known anthropologists such as Franz Boas (1858–1942) and Harold Conklin (1926–2016), who wrote the first dissertation on ethnoecology, studied traditional ecological taxonomic structures, and were important for bringing ecological wisdom possessed by other cultural groups into conversation with modern science. Meanwhile, the ethnobotanist Richard Schultes's (1915–2001) career spanned nearly five decades, and his research on the medical uses of previously unknown species revolutionized Western medicine (Schultes 1979, 1969); Schultes & Von Reis 1995). It was clear that Indigenous cultures possessed detailed knowledge of plants and animals that was foreign to Western science. Moreover, their knowledge was the product of empirical inquiry. Their perception of reality (ontology) and the ways in which they gathered knowledge about the world (epistemology) were empirical, data-driven. It was simply that the methodology was different from Western science, owning to those differential epistemologies and ontologies.

In some cases, those alternative ontologies and epistemologies were joined with Western scientific investigations to improve practical outcomes for Indigenous and traditional populations. Through these intercultural encounters new decision-making models emerged, and the reconciliation of disparate methods and data sets for common management goals was an important step toward preserving traditional knowledge and lifeways (Berkes 2018; Berkes et al. 1998). Encounters between Western and Indigenous science are not always uniformly positive (Baviskar 2005; Guha 2000, 2010; Shiva 2016). Nonetheless, simply the process of negotiating between the sciences of cultural, ethnic, and ethical others with some sort of empathy and vulnerability is a valuable development (Johnston 2014).

Becoming Kin Again

While some of the thinkers gathered here challenged Western scientific epistemologies, others worked to change them from within. For instance, Jane Goodall was a scientist who bucked scientific conventions, stimulating a significant shift in how biologists approached fieldwork. Moreover, rather than treating the chimpanzees she studied as objects designated by numbers, Goodall gave her chimps names and highlighted things that other scientists had not. She speculated about their emotional state and how they fostered intersubjective bonds between members of the troop. Goodall also noted that in some cases collective emotional displays were spurred by encounters with features of the environment. What Goodall termed waterfall dances were just one example where chimpanzees performed loud vocalizations and exhibited frenetic behavior. In the twenty-first century, Goodall even claimed that chimpanzees (and possibly other primates) exhibited some sort of spirituality (Goodall 2005; King 2007). This is one of those environmental religions that do not fall under the rubric of "world religions," but which are nonetheless important forms of religious cultural production. The famous conservation biologist Carl Safina is another who has written provocatively about what humans might learn from animals in terms of interspecies relations (Safina 2023), and animal cognition and emotion (Safina 2015, 2019, 2021, 2023).

There are also scholars who have advanced the notion of kinship with animals from within the world religions. So, for instance, the animal studies scholar Paul Waldau was instrumental in creating useful intellectual interventions among some religious groups, as well as writing extensively about animal ethics and law. His written corpus includes books on Buddhist and Christian views of animals, as well as the relationships between religions, science, and ethics (Waldau 2003, 2010, 2013; Waldau & Patton 2006). Laura Hobgood was influential in exploring Christian perspectives on animals. She provided historical overviews of animals in religious texts and practices, as well as provocative arguments about seeing the world through the eyes of animals (Hobgood-Oster 2003, 2010, 2014).

In many instances, the wisdom of TEK is informed by human-animal relationships. This was apparent in the New Animism, reviewed in Chapter 1, where humans approached the world and the other-than-human persons who dwelled therein with a certain etiquette (Harvey 2013; Cheney & Weston 1999). The scholar of religions Graham Harvey, one of the main exponents of this perspective, pointed to the groundbreaking work of the anthropologist Irving Hallowell (1892–1974). Hallowell's fieldwork among Ojibway populations explored Native attentiveness to animals, but also to ostensibly non-living entities like stones, or landscapes (Hallowell 1967, 1976). The anthropologist Richard Nelson investigated the importance of animal species (particularly corvids) to Native populations in Alaska (Nelson 1983). The anthropologist Mary Douglas's fieldwork provided one of the most famous examples of the religious importance of animals with her analysis of the pangolin in Lele society (Douglas 1966). Other examples are legion (see, e.g., Levi-Strauss 1969, 1979; Tambiah 1969). While animals have often played a secondary role in many of the world religions, they figure prominently in accounts from anthropologists and human ecologists (Shepard 1967, 1973, 1982, 1996, 1998). Many societies, both historically and today, maintain important relations with the non-human world through ritual. These represent visions of personhood that are not relegated only to humans, but to other-than-human animals (Harner 1990).

Taking the ethical sequence (to use Aldo Leopold's terms) beyond the animal to the plant realm extended the notion of personhood to plant regimes and even specific plant species. As noted above, the ethnobotanist Richard Schultes cataloged many medically important plant species in the Amazon (Plotkin 1993; Schultes & Von Reis 1995; Schultes & Farnsworth 1980). Oftentimes plant species are imagined to be agents. In other words, plants have human-like qualities and characteristics and are often imagined as appearing in humanoid form (Pinchbeck & Rokhlin 2021). This was especially common for plants that had important ritual functions. In some cases, landscapes are themselves imagined as alive, organismic, and agential (Kohn 2013).

These expanded notions of personhood with other animals, plants, and even ecosystems are some of the ways in which people are attempting to reimagine concepts of kinship. Imagined kinships that go beyond the biological relations that are typically considered kin have been important for the development of human societies. Imagined kinship has provided disparate groups of people with the perception of a common cause. For example, Muslims from all over the world who emigrated to Afghanistan to drive out Soviet and later American invaders had a sense of imagined kinship with those oppressed Afghans. Similar identity fusion was observed in conflicts between Israel and Palestine, and also among militia groups in the United States who find common cause in resisting governmental control of land or other resources (Quammen 2020, 2023). Sometimes, then, an expanded sense of kinship might foment intergroup conflict. However, it is also a crucial ingredient in peacebuilding. Recognizing the humanity, the dignity of cultural, ethnic, and ethical others is a prerequisite for pursuing a just and resilient future. It is important to be careful with our kinship commitments lest they lead toward exclusion or domination of some individuals or groups.

To conclude, I have reviewed some of the themes that were common to many of the thinkers collected here: ecological disruption caused by shifts in the mode of production; the subsequent emergence of anomie as a reaction to growing alienation of humans from each other and from the natural world; the importance of multiple epistemologies and ontologies in creating sustainable societies; and understandings of kinship and belongingness. Each of the key thinkers gathered here significantly shaped scholarship at the nexus of nature, culture, and associated religions. Further, each represents a different academic discipline that contains important lessons for how to investigate and analyze environmental religions. The academy was a product of the Enlightenment values of intergroup cooperation and tolerance, fostered by the spread of liberal political philosophy and policies. The liberal focus on intergroup cooperation and tolerance, however, has advanced in such a way that it invites to the table of international relations only those political entities that are already "domesticated," which at least tacitly play by the same rules. The result is the exclusion of strong religions, those that have powerful exclusivist principles and practices (Almond et al. 2003; Marty & Appleby 1997). When it comes to core values and deep beliefs, the model of political behavior that assumes "rational actors" is insufficient to account for a much more nuanced reality. Especially when it comes to sacred values, individuals and populations do not act rationally for their own, individual, short-term benefit (Atran 2011; Atran et al. 2005, 2014; Atran & Medin 2008). Are there types of kinship that can extend beyond the parochial intercultural differences that drive conflict? Might there be some basis for promoting a collective moral approach that reduces conflict and concomitant ecological destruction? Would such an expanded cultural kinship foster a more just global community? The scholar of religions Bron Taylor has advanced what he called a global, terrapolitan Earth ethic, wherein the global community is gradually increasingly motivated by widespread acceptance of Darwinian science and recognition of ongoing ecological disruption (Taylor 2010). The religion and nature scholar Whitney Bauman advocated for a planetary ethic which depends upon the celebration of diversity and difference, multiple perspectives, and multiple ways of knowing. These are promising developments. It remains to be seen, however, whether the coming decades will foster a greater openness to the moral importance of the environment or will witness the withering of global narratives of tolerance and cooperation. In any case, environmental religions will remain a prominent feature of both local and global environmental contestations. It is my hope that efforts at becoming kin again will bear fruit, with our imagined allegiances reaching out to include not just cultural, ethnic, and ethical others, but also species and ecosystem others.

REFERENCES

Abram, David. 1996. *The Spell of the Sensuous: Perception and Language in a More-than-Human World*. New York: Pantheon Books.

———. 2010. *Becoming Animal: An Earthly Cosmology*. New York: Pantheon Books.

Affairs, Berkley Center for Religion, Peace and World. 2001. "The Global Ethic: Hans Küng's Lasting Gift to the World." Accessed August 7, 2023. https://berkleycenter.georgetown.edu/responses/the-global-ethic-hans-kung-s-lasting-gift-to-the-world.

Alberts, Bruce, Alexander Johnson, Julian Lewis, Martin Raff, Keith Roberts, and Peter Walter. 2002. "Epidermis and Its Renewal by Stem Cells." In *Molecular Biology of the Cell*. 4th Edition. Garland Science. https://www.ncbi.nlm.nih.gov/books/NBK26865/.

Almond, Gabriel, Scott Appleby, and Emmanuel Sivan. 2023. *Strong Religion: The Rise of Fundamentalisms around the World*. Chicago: Univesity of Chicago Press.

Andelson, Robert. 1991. *Commons Without Tragedy: Protecting the Environment from Overpopulation–A New Approach*. London: Shepheard-Walwyn.

Arnal, William, and Russell McCutcheon. 2013. *The Sacred Is the Profane: The Political Nature of "Religion."* New York: Oxford University Press.

Atran, Scott. 2002. *In Gods We Trust: The Evolutionary Landscape of Religion*. New York: Oxford University Press.

Atran, Scott. 2011. *Talking to the Enemy Faith, Brotherhood, and the (Un)making of Terrorists*. New York Ecco Press.

Atran, Scott, and Douglas Medin. 2008. *The Native Mind and the Cultural Construction of Nature*. Cambridge, MA: MIT Press.

Atran, Scott, Douglas Medin, and Norbert Ross. 2005. "The Cultural Mind: Environmental Decision Making and Cultural Modeling Within and Across Populations." *Psychological Review* 112 (4): 744–776.

Atran, Scott, Douglas Medin, Norbert Ross, Elizabeth Lynch, Valentina Vapnarsky, Edilberto Ucan Ek', John Coley, Christopher Timura, and Michael Baran. 2002.

"Folkecology, Cultural Epidemiology, and the Spirit of the Commons: A Garden Experiment in the Maya Lowlands, 1991-2001." *Current Anthropology* 43 (3): 421–450.

Atran, Scott, Hammad Sheikh, and Angel Gomez. 2014. "For Cause and Comrade: Devoted Actors and Willingness to Fight." *Cliodynamics* 5 (1). https://doi.org/10.21237/C7clio5124900.

Axtell, James. 1992. *Beyond 1492: Encounters in Colonial North America*. New York: Oxford University Press.

Baker, Lauren, Samara Brock, Luisa Cortesi, Aysen Eren, Chris Hebdon, Francis Ludlow, Jeffrey Stoike, and Michael Dove. 2017. "Mainstreaming Morality: An Examination of Moral Ecologies as a Form of Resistance." *Journal for the Study of Religion, Nature and Culture* 11 (1): 23–55. https://doi.org/10.1558/jsrnc.27506.

Basso, Keith H. *Basso, Keith. 1996. Wisdom Sits in Places: Landscape and Language Among the Wester Apache*. Albuquerque: University of New Mexico Press

Bateson, Gregory. 1979. *Mind and Nature: A Necessary Unity*. New York: Dutton.

Bauman, Whitney. 2014. *Religion and Ecology: Developing a Planetary Ethic*. New York: Columbia University Press.

———. 2014. "The Myth of the Secular and the Shift Toward Secularisms." In *Science and Religion: One Planet, Many Possibilities*. New York: Routledge.

———. 2019. *Meaningful Flesh: Reflections on Religion and Nature for a Queer Planet*. Goleta, CA: Punctum Books.

Bauman, Whitney, Richard Bohannon, and Kevin J. O'Brien. 2011. *Inherited Land: The Changing Grounds of Religion and Ecology*. Eugene, OR.: Pickwick Publications.

Baxter, William F. 1974. *People or Penguins: The Case for Optimal Pollution*. New York: Columbia University Press.

Baviskar, Amita. 2005. *In the Belly of the River: Tribal Conflicts over Development in the Narmada Valley*. New York: Oxford University Press.

Bekoff, Mark. 2006. *Animal Passions and Beastly Virtues: Reflections on Redecorating Nature*. Philadelphia: Temple University Press.

Bellah, Robert N, and Steven M Tipton. 2006. "The New Religious Consciousness and the Crisis of Modernity." In *The Robert Bellah Reader*, 265–284. Durham: Duke University Press.

Bennett, Jane. 2010. *Vibrant Matter: A Political Ecology of Things*. Durham: Duke University Press.

Benthall, Jonathan. 2008. *Returning to Religion Why a Secular Age Is Haunted by Faith*. London; New York: I.B. Tauris.

Bering, Jesse, and David Bjorkland. 2004. "The Natural Emergence of Resoning About the Afterlife as a Developmental Regularity." *Developmental Psychology* 40 (2): 217–233.

Berkes, Fikret. 1999. *Sacred Ecology: Traditional Ecological Knowledge and Resource Management*. Philadelphia, PA: Taylor & Francis.

———. 2012. *Sacred Ecology*. New York: Routledge.

———. 2018. *Sacred Ecology*. 4th edition. New York: Routledge.

Berkes, Fikret, Johan Colding, and Carl Folke. 2009. *Navigating Social-Ecological Systems: Building Resilience for Complexity and Change*. Cambridge: Cambridge University Press.

Berry, Evan. 2015. *Devoted to Nature: The Religious Roots of American Environmentalism*. Oakland, California: University of California Press.

Berry, Thomas. 2000. *The Great Work: Our Way Into the Future*. New York: Crown Books.
Boff, Leonardo. 1997. *Cry of the Earth, Cry of the Poor*. Maryknoll, NY: Orbis Books.
Bohm, David. 1981. *Wholeness and the Implicate Order*. London; Boston: Routledge & Kegan Paul.
Bohm, David, and B. J. Hiley. 1993. *The Undivided Universe: An Ontological Interpretation of Quantum Theory*. London: Routledge.
Boulding, Kenneth. 1966. "The Economics of the Coming Spaceship Earth." In *Environmental Quality in a Growing Economy*, 3–14. Baltimore, MD: Resources for the Future/Johns Hopkins University Press.
Boyer, Pascal. 2001. *Religion Explained: The Evolutionary Origins of Religious Thought*. New York: Basic Books.
Brown, Dee. 1970. *Bury My Heart at Wounded Knee*. New York: Henry Holt and Company.
Bulbulia, Joseph. 2004. "The Cognitive and Evolutionary Psychology of Religion." *Biology and Philosophy* 19 (5): 655–86.
———. 2009. "Charismatic Signaling." *Journal for the Study of Religion, Nature & Culture* 3 (4): 518–551.
Bullard, Robert D. 1990. *Dumping in Dixie: Race, Class, and Environmental Quality*. Boulder: Westview Press.
Burhenn, Herbert. 1997. "Ecological Approaches to the Study of Religion." *Method & Theory in the Study of Religion* 9 (2): 111–26.
Burkert, Walter. 1996. *Creation of the Sacred: Tracks of Biology in Early Religions*. Cambridge, MA: Harvard University Press.
Cajete, Gregory. 2000. *Native Science: Natural Laws of Interdependence*. First edition. Santa Fe, New Mexico: Clear Light Publishers.
———. 2016. *Native Science: Natural Laws of Interdependence*. First Edition. Santa Fe, N.M: Clear Light Publishers.
———. 2015. *Indigenous Community: Rekindling the Teachings of the Seventh Fire*. Ist edition. St. Paul, Minnesota: Living Justice Press.
Callicott, J. Baird. 1980. "Animal Liberation: A Triangular Affair." *Environmental Ethics* 2 (Winter): 311–338.
Callicott, J. Baird. 1985. "Intrinsic Value, Quantum Theory, and Environmental Ethics." *Environmental Ethics*. Vol. 7 (3): 257–275.
———. 1994. *Earth's Insights: A Multicultural Survey of Ecological Ethics from the Mediterranean Basin to the Australian Outback*. Berkeley: University of California Press.
———2011. "The Worldview Concept and Aldo Leopold's Project of "World View" Remediation. *Journal for the Study of Religion, Nature, and Culture*. Vol. 5 (4): 510-528.
———2013. *Thinking Like a Planet: The Land Ethic and the Earth Ethic. New York*: Oxford University Press.
Callicott, J. Baird, and Michael P. Nelson, eds. 1998. *The Great New Wilderness Debate*. Athens, GA: University of Georgia Press.
Callicott, J. Baird, and Michael P Nelson. 2004. *American Indian Environmental Ethics: An Ojibwa Case Study*. Upper Saddle River, N.J.: Prentice Hall.
Capra, Fritjof. 1975. *The Tao of Physics: An Exploration of the Parallels between Modern Physics and Eastern Mysticism*. Berkeley: Shambhala.

———. 2002. *The Hidden Connections: Integrating the Biological, Cognitive, and Social Dimensions of Life into a Science of Sustainability*. New York: Doubleday.
Carp, Richard. 2012. "Toward a Resilient Academy." In *Higher Education for Sustainability: Cases, Challenges, and Opportunities Across the Curriculum*, 233–37. New York: Routledge.
Carson, Rachel. 1951. *The Sea around Us*. New York: Oxford University Press.
———. 1955. *The Edge of the Sea*. Boston: Houghton Mifflin.
Carson, Rachel. 1962. *Silent Spring*. Boston; Cambridge, MA: Houghton Mifflin; Riverside Press.
Castro, Eduardo Viveiros de. 1998. "Cosmological Deixis and Amerindian Perspectivism." *The Journal of the Royal Anthropoloigcal Institute of Great Britain and Ireland* 4 (3): 469–88.
Chalee, Toffany. "The Rights of Nature — Can an Ecosystem Bear Legal Rights?" 2021. *State of the Planet* (blog). Accessed April 22, 2021. https://news.climate.columbia.edu/2021/04/22/rights-of-nature-lawsuits/.
Chapple, Christopher Key. 2000. *Hinduism and Ecology: The Intersection of Earth, Sky, and Water*. Cambridge: Harvard University Press.
———. 2002. *Jainism and Ecology: Nonviolence in the Web of Life*. Cambridge: Harvard University Press.
Cheney, Jim, and Anthony Weston. 1999. "Environmental Ethics as Environmental Etiquette: Toward an Ethics-Based Epistemology." *Environmental Ethics* 21 (2): 115–34.
Chiao, Joan Y. 2011. "Cultural Neuroscience: Visualizing Culture-Gene Influences on Brain Function." In *The Oxford Handbook of Social Neuroscience*, edited by Jean Decety and John T. Cacioppo Oxford University Press.
———. 2011. *Visions of Discovery: New Light on Physics, Cosmology, and Consciousness*. Cambridge: Cambridge University Press.
Chidester, David. 2001. *Patterns of Action: Religion and Ethics in a Comparative Perspective*. Belmont, CA: Wadsworth/Thomson Learning.
———. 2003. "'Classify and Conquer'": Friedrich Max Muller, Indigenous Religious Traditions, and Imperial Comparative Religion." In *Beyond Primitivism: Indigenous Religious Traditions and Modernity*, 71–88. New York: Routledge.
———. 2005. *Authentic Fakes: Religion and Popular Culture in America*. Berkeley: University of California Press.
Chidester, David, and Edward Tabor Linenthal. 1995. *American Sacred Space*. Bloomington: Indiana University Press.
Christ, Carol. 1995. *Diving Deep and Resurfacing: Women Writers on the Spiritual Quest*. 3rd ed. Boston: Beacon Press.
———. 1997. *Rebirth of the Goddess: Finding Meaning in Feminist Theology*. New York: Routledge.
Charger, Harry & Ulrike Wiethaus. 2013. *The Seven Rites of the Lakota*. Create Space Independent Publishing Platform.
Clark, John. 2005a. "Anarchism." In *The Encyclopedia of Religion and Nature*. London: Continuum.
———. 2005b. "Social Ecology." In *The Encyclopedia of Religion and Nature*. London: Continuum. https://www.oxfordreference.com/display/10.1093/acref/9780199754670.001.0001/acref-9780199754670-e-848.

Colin Campbell. 1972. "The Cult, The Cultic Milieu and Secularization." *A Sociological Yearbook of Religion in Britain*, 119–136.
Committee on Restoration of the Greater Everglades Ecosystem, National Research Council. 2003. *Adaptive Monitoring and Assessment for the Comprehensive Everglades Restoration Plan*. The National Academies Press.
Cone, James. 1970. *A Black Theology of Liberation*. Philsadelphia: Lippincott.
Cone, James H. 1970. *A Black Theology of Liberation*. 1st edition. C. Eric Lincoln Series in Black Religion. Philadelphia: Lippincott.
Conklin, Beth and Laura Graham. 1995. "The Shifting Middle-Ground: Amazonian Indians and Eco-Politics." *American Anthropologist* 97 (4): 695–710.
Covey, Cyclone, trans. 1961. *Cabeza de Vaca's Adventures in the Unknown Interior of America*. New York: Collier Books.
Craddock, Elaine, and Vinay Lal. 2006. "Chipko Movement." In *The Encyclopedia of Religion and Nature*. London: Continuum. http://www.oxfordreference.com/display/10.1093/acref/9780199754670.001.0001/acref-9780199754670-e-156
Crocker, Jon. 1985. *Vital Souls: Bororo Cosmology, Natural Symbolism, and Shamanism*. Tucson: University of Arizona.
Cronin, William. 1995. *Uncommon Ground: Toward Reinventing Nature*. New York: W.W. Norton & Co.
Crosby, Alfred. 2003. *The Columbian Exchage: Biological and Cultural Consequences of 1492*. Durham, NC: Duke University Press.
Daly, Herman E. 1973. *Toward a Steady-State Economy*. San Francisco: W.H. Freeman.
———. 1996. *Beyond Growth: The Economics of Sustainable Development*. Boston: Beacon Press.
Daly, Herman E, John B Cobb, and Clifford W Cobb. 1989. *For the Common Good: Redirecting the Economy toward Community, the Environment, and a Sustainable Future*. Boston: Beacon Press.
Davies, Paul, ed. 1989. *The New Physics*. First Edition. Cambridge: Cambridge University Press.
Davies, P. C. W. 1990. *God and the New Physics*. London: Penguin.
Davison, Aidan. 2001. *Technology and the Contested Meanings of Sustainability*. Albany, N.Y.: State University of New York Press.
Deloria, Philip Joseph. 1998. *Playing Indian*. New Haven: Yale University Press.
Deloria, Vine. 1973. *God Is Red*. New York: Dell Publishing Company.
———. 1974. *Behind the Trail of Broken Treaties: An Indian Declaration of Independence*. New York: Dell Publishing.
———. 1985. *American Indian Policy in the Twentieth Century*. Norman: University of Oklahoma Press.
———. 1992. *God Is Red: A Native View of Religion*. 2nd edition. Golden, CO: Fulcrum Publishing.
———. 1994. *God Is Red: A Native View of Religion*. Golden, CO: Fulcrum Publishing.
———. 1997. *Red Earth White Lies: Native Ameircans and the Myth of Scientific Fact*. Wheat Ridge, CO: Fulcrum Press.
———. 1999. *For This Land: Writings on Religion in America*. New York: Routledge.
———. 2000. Review of *The Speculations of Krech, The Ecological Indian, by Shepard Krech*. *Worldviews* 4 (3): 283–293.

———. 2003. *God Is Red: A Native View of Religion*, 30th anniversary edition. Golden, United States: Fulcrum Publishing.

———. 2006. *The World We Used to Live In: Remembering the Powers of the Medicine Men*. Golden, United States: Fulcrum Publishing.

Denevan, William M. 1992. "The Pristine Myth: The Landscape of the Americas in 1492." *Annals of the Association of American Geographers* 82 (3): 369–385.

Denevan, William. 2011. "The 'Pristine Myth' Revisited." *Geographical Review* 101 (4): 576–591.

De Waal, F. B. M. de. 1996. *Good Natured: The Origins of Right and Wrong in Humans and Other Animals*. Cambridge, Mass.: Harvard University Press.

———. 2009. *The Age of Empathy: Nature's Lessons for a Kinder Society*. New York: Harmony Books.

———. 2014. "The Bonobo and the Atheist: In Search of Humanism Among the Primates." Nature, vol. 508, no. 7495, 184.

———. 2020. *Mama's Last Hug: Animal Emotions and What They Teach Us About Ourselves*. London: Granta.

Diamond, Jared M. 2011. *Collapse: How Societies Choose to Fail or Succeed*. New York, NY: Penguin Books.

Dioszegi, V., and Mihaly Hoppal. 1995. *Shamanism in Siberia*. Reprint edition. Budapest: Akadémiai Kiadó.

———. 1996. *Folk Beliefs and Shamanic Traditions in Siberia*. Budapest: Akadémiai Kiadó.

———. 1998. *Shamanism: Selected Writings. Budapest*. Budapest: Akadémiai Kiadó.

Douglas, Mary. 1966. *Purity and Danger: An Analysis of Concepts of Pollution and Taboo*. New York: Frederick A. Praeger.

———. 1973. *Natural Symbols: Explorations in Cosmology*. New York: Vintage Books.

———. 1975. *Implicit Meanings: Essays in Anthropology*. London: Routledge & Kegan Paul.

Dubuisson, Daniel. 2003. *The Western Construction of Religion: Myths, Knowledge, and Ideology*. Baltimore, Md.: Johns Hopkins University Press.

Durkheim, Emile. 1965. *The Elementary Forms of Religious Life*. New York: The Free Press.

———. 1972. *Selected Writings*. Edited by Anthony Giddens. Cambridge: Cambridge University Press.

Eckersley, Robyn. 1998. "The Death of Nature and the Birth of the Ecological Humanities." *Organization & Environment* 11 (2): 183–185.

Eisenberg, Evan. 1998. *The Ecology of Eden*. London: Picador.

Eliade, Mircea. 1959. *The Sacred and the Profane: The Nature of Religion*. New York: Harcourt.

———. 1978. *A History of Religious Ideas*. Chicago: University of Chicago Press.

Elliot, Robert. 1997. *Faking Nature: The Ethics of Environmental Restoration*. London; New York: Routledge.

Epstein, Yaffa, Aaron M. Ellison, Hugo Echeverría, and Jessica K. Abbott. 2023. "Science and the Legal Rights of Nature." *Science* 380 (6646): eadf4155. https://doi.org/10.1126/science.adf4155.

Entine, Jon. 2014. "Vandana Shiva, Anti-GMO Celebrity: 'Eco Goddess' Or Dangerous Fabulist?" *Forbes*. Accessed June 9, 2023. https://www.

forbes.com/sites/jonentine/2014/01/29/vandana-shiva-anti-gmo-celebrity-eco-goddess-or-dangerous-fabulist/.
Feuerbach, Ludwig. 1957. *The Essence of Christianity*. Translated by George Eliot. New York: Harper.
Fields, Rick. 2012. *How the Swans Came to the Lake: A Narrative History of Buddhism in America*. Boston: Shambhala Publications.
Fitzgerald, Timothy. 2003a. *The Ideology of Religious Studies*. New York: Oxford University Press.
———. 2003b. "Playing Language Games and Performing Rituals: Religious Studies as Ideological State Apparatus." *Method & Theory in the Study of Religion* 15 (3): 209–254.
———. 2015. "Critical Religion and Critical Research on Religion: Religion and Politics as Modern Fictions." *Critical Research on Religion* 3 (3): 303–319.
Foltz, Richard, Frederick Mathewson. Denny, and Azizan Haji Baharuddin. 2003. *Islam and Ecology: A Bestowed Trust*. Cambridge, Mass.: Center for the Study of World Religions, Harvard University Press.
Frazer, James George. 2012. *The Golden Bough. Volume 2: A Study in Comparative Religion*. 1st edition. Cambridge Library Collection. New York: Cambridge University Press.
Freud, Sigmund. 1964. *The Future of an Illusion*. London: Hogarth Press.
———. 1998. *Totem and Taboo*. Minneola, NY: Dover Publications.
Gill, Sam D. 1987. *Mother Earth: An American Story*. Chicago: University of Chicago Press.
Gillespie, Michele, and Robert Beachy. 2007. *Pious Pursuits: German Moravians in the Atlantic World*. European Expansion and Global Interaction. New York: Berghahn Books.
Girardot, N. J., James Miller, and Xiaogan Liu. 2001. *Daoism and Ecology: Ways within a Cosmic Landscape*. Religions of the World and Ecology. Cambridge, MA: Center for the Study of World Religions, Harvard University Press.
Glacken, Clarence J. 1967. *Traces on the Rhodian Shore; Nature and Culture in Western Thought from Ancient Times to the End of the Eighteenth Century*. Berkeley: University of California Press.
Glass, Jeffrey. 2005. "Law, Religion, and Native American Lands." In *The Encyclopedia of Religion and Nature*, edited by Bron Raymond Taylor, 991–1000. London: Continuum.
Goodall, Jane. 2005. "Primate Spirituality." In *Encyclopedia of Religion and Nature*, edited by Bron Taylor, 1303–1306. London: Continuum.
Goodall, Jane, and Marc Bekoff. 2014. *The Ten Trusts: What We Must Do to Care for the Animals We Love*. San Francisco: Harper One.
Greenwood, Susan. 2022. "A Spectrum of Magical Consciousness: Conspiracy Theories and the Stories We Tell Ourselves." *Anthropology Today* 38 (1): 3–7.
Grim, John, and Mary Evelyn Tucker. 2014. *Ecology and Religion*. New York: Routledge.
Guha, Ramachandra. 1989. "Radical American Environmentalism and Wilderness Preservation: A Third World Critique." *Environmental Ethics* 11 (1): 71–83.
———. 1995. "Radical American Environmentalism and Wilderness Preservation: A Third World Critique." In *Survey of Ecological Economics*, edited by Rajaram Krishnan, Jonathan M. Harris, and Neva R. Goodwin, 317-320.

———. 1996. “Two Phases of American Environmentalism: A Critical History.” In *Decolonizing Knowledge: From Development to Dialogue*, edited by Frédérique Apffel-Marglin and Stephen A. Marglin.
———. 2000. *Environmentalism: A Global History*. Longman World History Series. New York: Longman.
———. 2010. *The Unquiet Woods: Ecological Change and Peasant Resistance in the Himalaya*. 20th anniversary edition. Ranikhet, Bangalore: Permanent Black; Distributed by Orient Blackswan.
Gustafson, James M. 1996. *A Sense of the Divine: The Natural Environment from a Theocentric Perspective*. Cleveland, Ohio: Pilgrim Press.
Guthrie, Stewart. 1995. *Faces in the Clouds: A New Theory of Religion*. New York: Oxford University Press.
Gutiérrez, Gustavo. 2023. *A Theology of Liberation: History, Politics, and Salvation*. 50th anniversary edition. Maryknoll, NY: Orbis Books.
Hallowell, A. Irving. 1967. *Culture and Experience, A. Irving Hallowell*. Philadelphia Anthropological Society. Philadelphia: University of Pennsylvania Press.
———. 1976. *Contributions to Anthropology: Selected Papers of A. Irving Hallowell*. Chicago: University of Chicago Press.
Hand, Floyd “Looks for Buffalo.” (1998). *Learning Journey on the Red Road*. Pine Ridge, SD: Learning Journey Communications.
Hardin, Garrett. 1968. “The Tragedy of the Commons.” *Science* 162 (3859): 1243–1248.
———. 1973. *Stalking the Wild Taboo*. Los Altos, CA: W. Kaufmann.
———. 1974. “Lifeboat Ethics: The Case Against Helping the Poor.” *Psychology Today* 8: 38–43.
Harding, Sandra. 1991. *Whose Science? Which Knowledge? Thinking from Women's Lives*. Ithaca, NY: Cornell University Press.
———. 2002. *Is Science Multicultural? Postcolonialisms, Feminisms, and Epistemologies*. Bloomington: Indiana University Press.
———. 2015. *Objectivity and Diversity: Another Logic of Scientific Research*. Chicago: University of Chicago Press.
Harner, Michael J. 1990. *The Way of the Shaman*. 10th anniversary edition First Harper & Row paperback edition. San Francisco: Harper & Row.
Harris, Marvin. 1966. “The Cultural Ecology of India's Sacred Cattle.” *Current Anthropologoy* 7 (1): 51–54+55–66.
Harrod, Howard. 1992. *Renewing the World: Plains Indian Religion and Morality*. Tempe: University of Arizona Press.
———. 1995. *Beconing and Remaining a People: Native American Religions on the Northern Plains*. Tempe: University of Arizona.
Harvey, David. 2011. *A Brief History of Neoliberalism*. Reprinted. Oxford: Oxford Univ. Press.
Harvey, Graham. 2006. *Animism: Respecting the Living World*. New York: Columbia University Press.
———. 2013. *Food, Sex and Strangers: Understanding Religion as Everyday Life*. Durham, UK: Acumen Press.
Hart, John. 2005. “World Conference of Indigenous Peoples (Kari Oca, Brazil).” In *Encyclopedia of Religion and Nature*, edited by Bron Taylor, 2: 1763–65. London: Continuum.

Hatch, Nathan O. 1991. *Democratization of American Christianity*. New edition. New Haven, CT: Yale University Press.
Hessel, Dieter T., and Rosemary Radford Ruether. 2000. *Christianity and Ecology: Seeking the Well-Being of Earth and Humans*. Cambridge, MA: Harvard University Press, Harvard University Center for the Study of World Religions.
Hill, Jonathan David. 1996. *History, Power, and Identity: Ethnogenesis in the Americas, 1492-1992*. Iowa City, Iowa: University of Iowa Press.
Hobgood-Oster, Laura. 2008. *Holy Dogs and Asses: Animals in the Christian Tradition*. Urbana: University of Illinois Press.
———. 2014. *A Dog's History of the World: Canines and the Domestication of Humans*. Waco, Texas: Baylor University Press.
House, Freeman. 1999. *Totem Salmon: Life Lessons from Another Species*. Boston: Beacon Press.
Hukkinen, Janne. 2008. *Sustainability Networks: Cognitive Tools for Expert Collaboration in Social-Ecological Systems*. London: Routledge.
Hultkrantz, Åke. 1981. *The Religions of the American Indians*. Translated by Monica Setterwall.
———. 1997. *Native Religions of North America: The Power of Visions and Fertility*. Long Grove (Ill.): Waveland Press Inc.
Ivakhiv, Adrian. 2007. "Religion, Nature and Culture: Theorizing the Field." *Journal for the Study of Religion, Nature and Culture* 1 (1): 47–57.
James, George Alfred. 2013. *Ecology Is Permanent Economy: The Activism and Environmental Philosophy of Sunderlal Bahuguna*. Albany, NY: SUNY Press.
Johnston, Lucas F. (ed). 2013. *Higher Education for Sustainability: Cases, Challenges, and Opportunities from across the Curriculum*. New York, NY: Routledge.
———. 2014. "Sustainability as a Global Faith? The Religious Dimensions of Sustainability and Personal Risk." *Journal of the American Academy of Religion* 82 (1): 47–69.
———. 2016. "Cultivating an Academy We Can Live With: The Humanities and Education for Sustainability." *Religions* 7 (10): 120. https://doi.org/10.3390/rel7100120.
Johnston, Lucas F., and Bron Taylor. 2016. "Trends in Religion and Environmental Politics into the Twenty-First Century and Beyond." In *The Wiley Blackwell Companion to Politics and Religion in America*. Chichester, UK: Wiley-Blackwell.
Johnson, Greg, and Siv-Ellen Kraft. 2017. "Standing on the Sacred: Ceremony, Discourse, and Resistance in the Fight against the Black Snake." *Journal for the Study of Religion, Nature and Culture* 11 (1): 131–147. https://doi.org/10.1558/jsrnc.32617.
Johnson, Mark. 1993. *Moral Imagination: Implications of Cognitive Science for Ethics*. Chicago: University of Chicago Press. http://catdir.loc.gov/catdir/enhancements/fy0608/92029896-t.html.
Johnston, Lucas F. 2006. "The 'Nature' of Buddhism: A Survey of Relevant Literature and Themes." *Worldviews: Environment, Culture, Religion* 10(1): 69–99.
———. 2013. *Religion and Sustainability: Social Movements and the Politics of the Environment*. Sheffield; Bristol, CT: Equinox.

———. 2016. "Dancing My Prayers: Material Culture and Spiritual Practices of Improvisational Rock Music Subcultures." In *Practical Spiritualities in the Media Age*, edited by Monica Emerich and Curtis Coats. New York: Bloomsbury Press.

———. 2018. "Worldview Analysis in Comparative Context: Fishing for Data in Muddy Waters." *Religions* 9 (9): 261–279.

Johnston, Lucas, and Dave Aftandilian, eds. 2018. *Grounding Education in Environmental Humanities: Place-Based Pedagogies in the South*. New York: Routledge.

Johnston, Lucas F, and Bauman, Whitney. 2014. *Science and Religion: One Planet, Many Possibilities*.

Johnston, Lucas, and Wall Robert. 2014. "The Camel and the Eye of the Needle: Religion, Moral Exchange, and Social Impacts." In The Changing World Religions Map, edited by Stanley Brunn, 1259–1274. New York: Springer.

Johnston, Lucas F., and Bron Taylor. 2016. "Trends in Religion and Environmental Politics into the Twenty-First Century and Beyond." In *The Wiley Blackwell Companion to Politics and Religion in America*, 454–469. Chichester, UK: Wiley-Blackwell.

Jolly, Richard. 2004. *UN Contributions to Development Thinking and Practice*. Bloomington: Indiana University Press.

Jordan, William R. 2003. *The Sunflower Forest: Ecological Restoration and the New Communion with Nature*. Berkeley: University of California Press.

———. 2005. "Restoration Ecology and Ritual." In *Encyclopedia of Religion and Nature*, edited by Bron Taylor, 1379–81. London: Continuum.

Jordan III, William, and George Lubick. 2011. *Making Nature Whole: A History of Ecological Restoration*. Washington, DC: Island Press.

Jung, Carl. 1964. *Man and His Symbols*. New York: Dell Books.

Kalland, Arne. 2006. "The Religious Environmentalist Paradigm." In *The Encyclopedia of Religion and Nature*, London: Continuum.

Katz, Eric. 1992. "The Big Lie: Human Restoration of Nature." *Research in Philosophy and Technology*. Vol. 12: 231–241.

Katz, Eric. 2003. "The Big Lie: Human Restoration of Nature." In *Environmental Ethics: An Anthology*, edited by Andrew Light and Holmes Rolston III, 390–398. Malden, MA: Blackwell.

Katz, Eric. 2012. "Further Adventures in the Case against Restoration." *Environmental Ethics* 34 (1): 67.

Kearns, Laurel & Whitney Bauman. 2024. *Religion and Nature in North America: An Introduction*. London: Bloomsbury Press.

Kellert, Stephen R. 2006. "Biophilia." In *The Encyclopedia of Religion and Nature*. Continuum. https://www.oxfordreference.com/display/10.1093/acref/9780199754670.001.0001/acref-9780199754670-e-89

———. 2007. "Connecting with Creation: The Convergence of Nature, Religion, Science and Culture." *Journal for the Study of Religion, Nature and Culture* 1 (1): 25–37.

Kellert, Stephen R., and Timothy J. Farnham (eds). 2002. *The Good in Nature and Humanity: Connecting Science, Religion, and Spirituality with the Natural World*. Washington, DC: Island Press.

Kellert, Stephen, and E.O. Wilson, eds. 1993. *The Biophilia Hypothesis*. Washington, DC: Island Press.

King, Barbara J. 2007. *Evolving God: A Provocative View of the Origins of Religion*. New York: Doubleday.

King, Jr., Luther King, Jr. 1963. "'Letter from Birmingham Jail,' The Martin Luther King, Jr. Research and Education Institute." Accessed September 6, 2023. https://kinginstitute.stanford.edu/letter-birmingham-jail.

Kiper, Joseph, and Richard Sosis. 2014. "Moral Intuitions of the Religious System: An Adaptationist Account." *Philosophy, Theology, and the Sciences* 1 (2): 172–99.

Klare, Michael T. 2008. *Rising Powers, Shrinking Planet: The New Geopolitics of Energy*. New York: Metropolitan Books.

———. 2012. *The Race for What's Left: The Global Scramble for the World's Last Resources*. New York: Metropolitan Books.

Kohn, Eduardo. 2013. *How Forests Think: Toward an Anthropology Beyond the Human*. Berkeley: University of California Press.

Krech, Shepard. 1999. *The Ecological Indian: Myth and History*. New York: W.W. Norton & Co.

Kant, Immanuel. 1881. *Immanuel Kant's Critique of Pure Reason. Vol 1*. London: Macmillan.

Küng, Hans. 1998. *A Global Ethic for Global Politics and Economics*. New York: Oxford University Press.

Laack, Isabel. 2020. "The New Animism and Its Challenges to the Study of Religion." *Method & Theory in the Study of Religion* 32 (2): 115–147.

Labate, Beatriz, and Clancy Cavnar, eds. 2018. *Plant Medicines, Healing and Psychedelic Science: Cultural Perspectives*. 1st edition. New York: Springer.

Labate, Beatriz, and Cavnar Caiuby. 2014. *Ayahuasca Shamanism in the Amazon and Beyond*. Oxford: Oxford University Press.

Lal, Vinay. 2005. "Chipko Movement." In *The Encyclopedia of Religion and Nature*. London: Continuum.

Lansing, John Stephen. 1991. *Priests and Programmers: Technologies of Power in the Engineered Landscape of Bali*. Princeton, NJ: Princeton University Press.

Leopold, Aldo Starker. 1949. *A Sand County Almanac and Sketches Here and There. Illustrated by Charles W. Schwartz*. Oxford University Press: New York.

Leopold, Aldo. 1966. *A Sand County Almanac. With Other Essays on Conservation from Round River*. New York: Oxford University Press.

Leopold, Aldo. 1991. *The River of the Mother of God: And Other Essays*. Edited by Susan Flader and J. Baird Callicott. Madison: University of Wisconsin Press. https://uwpress.wisc.edu/books/0316.htm.

LeVasseur, Todd. 2021. *Climate Change, Religion, and Our Bodily Future*. Lanham, Maryland: Lexington Books, an imprint of The Rowman & Littlefield Publishing Group, Inc.

Lévi-Strauss, Claude. 1969. *The Savage Mind*. English tr. Nature of Human Society Series. Chicago: The University of Chicago Press.

Levi-Strauss, Claude. 1979. *Myth and Meaning: Cracking the Code of Culture by Claude Levi-Strauss; Foreword by Wendy Doniger*. New York: Schocken Books, Distributed by Pantheon Books.

Levitin, Daniel J. 2006. *This Is Your Brain on Music: The Science of a Human Obsession*. New York, N.Y.: Dutton.

———. 2008. *The World in Six Songs: How the Musical Brain Created Human Nature*. New York: Dutton Books.

Light, Andrew, and Eric S. Higgs. 1996. "The Politics of Ecological Restoration." *Environmental Ethics*, Vol 28 (3): 227–247. 1996.

Lovelock, James. 1979. *Gaia: A New Look at Life on Earth*. Oxford: Oxford University Press.
Loy, David. 1997. "The Religion of the Market." *Journal of the American Academy of Religion* 65 (2): 275–90.
Maathai, Wangari. 2006. *Unbowed: A Memoir*. New York: Knopf.
———. 2010. *Replenishing the Earth: Spiritual Values for Healing Oursevles and the World*. The Crown Press.
Macy, Joanna. 2005. "The Council of All Beings." In *The Encyclopedia of Religion and Nature*. London: Continuum.
Malinowski, Bronislaw. 2014. *Magic, Science and Religion and Other Essays*. Chicago: McCormick Press.
———. 2020. *Argonauts of the Western Pacific*. Norwich: Outlook Verlag.
Malthus, Thomas, and Geoffrey Gilbert. 2008. *An Essay on the Principle of Population*. Oxford World's Classics. Oxford, New York: Oxford University Press.
Marsh, George Perkins. 1864. *Man and Nature*. Cambridge: Belknap Press of Harvard University Press.
Martin, Craig. 2023. *Critical Introduction to the Study of Religion*. New York: Routledge.
Martin, Craig, and Russell McCutcheon, eds. 2012. *Religious Experience: A Reader*. New York: Routledge.
Martinez, David. 2019. *Life of the Indigenous Mind: Vine Deloria Jr. and the Birth of the Red Power Movement*. New Visions in Native American and Indigenous Studies. Lincoln, NE: University of Nebraska Press.
Marty, Martin E., and R. Scott Appleby, eds. 1997. *Religion, Ethnicity, and Self-Identity: Nations in Turmoil*. First Edition. Hanover: Univ Press of New England.
Masuzawa, Tomoko. 2005. *The Invention of World Religions, or, How European Universalism Was Preserved in the Language of Pluralism*. Chicago: University of Chicago Press.
Mburu, G. 2005. "Kenya Green Belt Movement." In *Encyclopedia of Religion and Nature*, edited by Bron Taylor. London: Continuum Press.
McCoy, Kate, Eve Tuck, and Marcia McKenzie, eds. 2017. *Land Education: Rethinking Pedagogies of Place from Indigenous, Postcolonial, and Decolonizing Perspectives*. London: Routledge.
McCutcheon, Russell T. 1997. *Manufacturing Religion: The Discourse on Sui Generis Religion and the Politics of Nostalgia*. New York: Oxford University Press.
———. 2001. *Critics Not Caretakers: Redescribing the Public Study of Religion*. Albany: State University of New York Press.
———. 2003. *Manufacturing Religion: The Discourse on Sui Generis Religion and the Politics of Nostalgia*. Oxford: Oxford University Press.
———. 2005. *Religion and the Domestication of Dissent, or, How to Live in a Less than Perfect Nation*. London; Oakville, CT: Equinox Publishing.
———. 2007. *Studying Religion: An Introduction*. London; Oakville, CT: Equinox Pub.
———. 2019. *Studying Religion: An Introduction*. 2nd edition. London: Routledge. https://search.ebscohost.com/login.aspx?direct=true&scope=site&db=nlebk&db=nlabk&AN=1936884.
McFague, Sallie. 1994. *The Body of God: An Ecological Theology*. 3rd printing. Philadelphia: Fortress Press.
McGinnis, Michael. 1999. *Bioregionalism*. New York: Routledge.

McKay, Barbara. 2005. "Carson, Rachel." In *Encyclopedia of Religion and Nature*. Edited by Bron Taylor. London: Continuum.
Meadows, Donella H and Club of Rome. 1972. *The Limits to Growth; a Report for the Club of Rome's Project on the Predicament of Mankind*. New York: Universe Books.
Medin, Douglas L., and Megan Bang. 2014. *Who's Asking?: Native Science, Western Science, and Science Education*. Cambridge, Massachusetts: The MIT Press.
Meine, Curt. 2010. *Aldo Leopold His Life and Work*. Madison, WI: University of Wisconsin Press.
Merchant, Carolyn. 1980. *The Death of Nature: Women, Ecology, and the Scientific Revolution*. San Francisco: Harper & Row.
Merchant, Carolyn. 1989. *Ecological Revolutions: Nature, Gender, and Science in New England* Chapel Hill, NC: University of North Carolina Press.
Merchant, Carolyn. 1992. *Radical Ecology: The Search for a Livable World*. [Pbk. ed.]. Revolutionary Thought/Radical Movements. New York: Routledge.
———. 2007. *American Environmental History: An Introduction*. New York, NY: Columbia University Press.
———. 2013. *Reinventing Eden: The Fate of Nature in Western Culture*. London: Routledge.
———. 2015. *Autonomous Nature: Problems of Prediction and Control From Ancient Times to the Scientific Revolution*. New York: Taylor and Francis.
———. 2016. *Earthcare: Women and the Environment*. New York: Routledge.
Merchant, Carolyn. 2019. *After the Death of Nature: Carolyn Merchant and the Future of Human-Nature Relations*. New York: Routledge.
———. 2020. *The Anthropocene and the Humanities: From Climate Change to a New Age of Sustainability*. New Haven: Yale University Press.
Midgley, Mary. 1985. *Evolution as a Religion*. New York: Taylor & Francis.
Miles, Margaret, and Vandana Shiva. 1993. *Ecofeminism*. London: Zed Books.
Mithen, Steven. 1996. *The Prehistory of the Mind: A Search for the Origins of Art, Religion, and Science*. London: Thames and Hudson.
Mithen, Steven J. 2006. *The Singing Neanderthals: The Origins of Music, Language, Mind, and Body*. Cambridge: Harvard University Press.
Muir, John. 1997. *Nature Writings*. New York: Penguin.
Müller, F. Max. 2002. *Natural Religion; the Gifford Lectures Delivered before the University of Glasgow in 1888*. London, New York: Longmans, Green, and Co.
Müller, Max. 2002. *The Max Müller Reader*. Edited by John Stone. New York: Basingstoke: Palgrave.
Muraresku, Brian. 2020. *The Immortality Key: The Secret History of the Religion with No Name*. First edition. New York: St. Martin's Press.
Nabhan, Gary Paul. 1997. "Cultural Parallax: The Wilderness Concept in Crisis." In *Cultures of Habitat: On Nature, Culture, and Story*, 152–65. Washington, D.C., Emeryville, CA: Counterpoint, Distributed by Publishers Group West.
Nabokov, Peter. 2005. *Where the Lightning Strikes: The Lives of American Indian Sacred Places*. New York: Viking Press.
Naess, Arne. 1973. "The Shallow and the Deep, Long-range Ecology Movement. A Summary." *Inquiry* 16 (1–4): 95–100.
Nash, Roderick. 1973. *Wilderness and the American Mind*. New Haven: Yale University Press.Nelson, Richard K. 1983. *Make Prayers to the Raven: A Koyukon View of the Northern Forest*. Chicago: University of Chicago Press.

Newsom, Carol. 2000. "Common Ground: An Ecological Reading of Genesis 2-3." In *The Earth Story in Genesis*, edited by Norman C Habel, 60–72. Sheffield: Sheffield Academic.

Niehardt, John Gneisenau. 1972. *Black Elk Speaks; Being the Life Story of a Holy Man of the Oglala Sioux as Told to John G. Neihardt (Flaming Rainbow)*. London: Barrie & Jenkins.

Northcutt, Patsy and David Kennard, dir. The Journey of the Universe. Samos, Greece: Journey of the Universe, 2011, Film.

Norton, Bryan G. 2005. *Sustainability: A Philosophy of Adaptive Ecosystem Management*. Chicago: Univ. of Chicago Press.

Núñez Cabeza de Vaca, Alvar, and Cyclone Covey. 2003. *The Journey and Ordeal of Cabeza de Vaca: His Account of the Disastrous First European Exploration of the American Southwest*. Dover Books on Travel, Adventure. Mineola, NY: Dover Publications.

Odum, Howard T. 1970. *Environment, Power, and Society*. New York: Wiley-Interscience.

Odum, Howard T, and Elisabeth C Odum. 1976. *Energy Basis for Man and Nature*. New York: McGraw-Hill.

———. 2001. *A Prosperous Way down: Principles and Policies*. Boulder, CO: University Press of Colorado.

Oelschlager, Max. 1994. *Caring for Creation: An Ecumenical Approach to the Environmental Crisis*. New Haven: Yale University Press.

Ophuls, William. 1977. *Ecology and the Politics of Scarcity: Prologue to a Political Theory of the Steady State*. Paperback edition. San Francisco: W.H. Freeman.

Ostrum, Elinor. 1990. *Governing the Commons: The Evolution of Instiitutions for Collective Action*. New York: Cambridge Univesity Press.

———. 1994. *Rules, Games, and Common Pool Resources*. Ann Arbor: University of Michigan.

Otto, Rudolph. 1925. *The Idea of the Holy*. 2nd edition. London: Oxford University Press.

Ouderkirk, Wayne, and Jim Hill, eds. 2002. *Land, Value, Community: Callicott and Environmental Philosophy*. SUNY Series in Environmental Philosophy and Ethics. Albany: State University of New York Press.

Ouspensky, P.D. 1949. *Tertium Organum: The Third Cannon of Thought-A Key to the Enigmas of the World*. Translated by Nicholas Bessarboroff and Claude Bragdon. London: Routledge and Kegan Paul.

Palmer, Martin and Victoria Finlay. 2003 *Faith in Conservation: New Approaches to Religions and the Environment*. Washington, DC: World Bank Press.

Paget-Clarke, Nic & Vandana Shiva. 2004. *Corporate Biotechnology: A New Form of Colonization: An Interview with Dr. Vandana Shiva*. Lake Worth, FL: Downward Mobility.

Patton, Kimberley C., and Benjamin C. Ray, eds. 2000. *A Magic Still Dwells: Comparative Religion in the Postmodern Age*. Berkeley, CA: University of California Press.

Peterson, Anna Lisa, and Todd LeVasseur, eds. 2017. *Religion and Ecological Crisis: The "Lynn White Thesis" at Fifty*. Routledge Studies in Religion. New York: Routledge.

Pinchbeck, Daniel, and Sophia Rokhlin. 2021. *When Plants Dream: Ayahuasca, Amazonian Shamanism and the Global Psychedelic Renaissance*. London: Watkins Publishing.

Plotkin, Mark J. 1993. *Tales of a Shaman's Apprentice: An Ethnobotanist Searches for New Medicines in the Amazon Rain Forest.* New York: Viking.
Plumwood, Val. 1995. "Human Vulnerability and the Experience of Being Prey." *Quadrant* 39 (3): 29–34.
Proctor, James D. 2005. *Science, Religion, and the Human Experience. Oxford ; New* York: Oxford University Press.
Quammen, Betsy Gaines. 2020. *American Zion: Cliven Bundy, God & Public Lands in the West.* First Torrey House Press edition. Salt Lake City: Torrey House Press.
Quammen, Betsy Gaines. 2023. *True West: Myth and Mending on the Far Side of America.* First Torrey House Press edition. Salt Lake City: Torrey House Press.
Quinn, Daniel. 1996. *The Story of B.* New York: Bantam Books.
Rappaport, Roy Abraham. 1966. "Ritual in the Ecology of a New Guinea People: An Anthropological Study of the Tsembaga Maring." New York: Columbia University.
Rappaport, Roy A. 1999. *Ritual and Religion in the Making of Humanity.* Cambridge, UK; New York: Cambridge University Press.
Robin, Libby, Sverker Sörlin, and Paul Warde, eds. 2013. *The Future of Nature: Documents of Global Change.* Illustrated edition. New Haven: Yale University Press.
Rist, Gilbert. 1997. *The History of Development: From Western Origins to Global Faith.* London; New York: Zed Books.
Roof, Wade Clark. 1994. *A Generation of Seekers: The Spiritual Journeys of the Baby Boom Generation.* HarperCollins pbk ed. San Francisco: HarperSanFrancisco.
Ruether, Rosemary Radford. 1970. *The Radical Kingdom: the Western Experience of Messianic Hope.* New York: Paulist Press.
———. 1972. *Liberation Theology: Human Hope Confronts Christian History and American Power.* New York: Paulist Press.
———. 1974. *Religion and Sexism: Images of Women in the Jewish and Christian Traditions.* New York: Simon and Schuster.
———. 1977. *Mary, the Feminine Face of the Church.* Philadelphia: Westminster Press.
———. 1983. *Sexism and God-talk: Toward a Feminist Theology.* Boston: Beacon Press.
———. 1988. *Women-Church: Theologies and Practice of Feminist Liturgical Communities.* San Francisco: Harper & Row.
———. 1992. *God and Gaia: An Ecofeminst Theology of Earth Healing.* San Francisco: Harper San Francisco.
———. 1996a. *Womanguides: Readings toward a Feminist Theology.* Boston: Beacon Press.
———, ed. 1996b. *Women Healing Earth: Third World Women on Ecology, Feminism, and Religion.* Ecology and Justice. Maryknoll, N.Y.: Orbis Books.
———. 2002. *Women Healing Earth: Third World Women on Ecology, Feminism, and Religion.* Maryknoll, NY: Orbis Books.
———. 2005. *Integrating Ecofeminism, Globalization, and World Religions.* Lanham: Rowman and Littlefield.
———. 2007. *America, Amerikkka: Elect Nation and Imperial Violence.* Sheffield, UK: Equinox Press.
Ruether, Rosemary Radford, and Marc Ellis. 1990. *Beyond Occupation: American Jewish, Christian, and Palestinian Voices for Peace.* Boston: Beacon Press.

Safina, Carl: *What Are Animals Thinking and Feeling?* 2015. TEDTalks. http://www.ted.com/talks/view/id/2373

Safina, Carl. 2019. *Beyond Words: What Elephants and Whales Think and Feel.* New York: Roaring Brook Press.

Safina, Carl. 2021. *Becoming Wild: How Animal Cultures Raise Families, Create Beauty, and Achieve Peace.* First Picador paperback edition. New York: Picador, Henry Holt and Company.

———. 2023. *Alfie & Me: What Owls Know, What Humans Believe.* First edition. New York: W. W. Norton & Company.

Sale, Kirkpatrick. 1991. *The Conquest of Paradise: Christopher Columbus and the Columbian Legacy.* New York: Plume.

Saler, Benson. 1993. *Conceptualizing Religion: Immanent Anthropologists, Transcendent Natives, and Unbounded Categories.* New York: Berghahn Books.

———. 2004. "Towards a Realistic and Relevant 'Science of Religion.'" *Method and Theory in the Study of Religion* 16 (3): 205–23.

Sands, Robert R, and Linda R Sands. 2009. "Running Deep: Speculations on the Evolution of Running and Spirituality in the Genus Homo." *Journal for the Study of Religion, Nature and Culture* 3 (4): 552–77.

Santmire, H. Paul. 1970. *Brother Earth: Nature, God, and Ecology in Time of Crisis.* New York: T. Nelson.

Savage, Patrick E., Steven Brown, Emi Sakai, and Thomas E. Currie. 2015. "Statistical Universals Reveal the Structures and Functions of Human Music." *Proceedings of the National Academy of Sciences* 112 (29): 8987–8992.

Schaeffer, Francis A. 1970. *Pollution and the Death of Man: The Christian View of Ecology.* Wheaton, Ill.: Tyndale House Publishers.

Schneider-Meyerson, Matthew. 2015. *Peak Oil and Apocalyptic Environmentalism and Libertarian Political Culture.* Chicago: University of Chicago Press.

Schneider, Ralf, and Marcus Hartner. 2012. *Blending and the Study of Narrative: Approaches and Applications.* Berlin: De Gruyter.

Schultes, Richard Evans. 1969. "Hallucinogens of Plant Origin." *Science* 163 (3864): 245–254.

Schultes, Richard Evans, and Norman R. Farnsworth. 1980. "Ethnomedical, Botanical and Phytochemical Aspects of Natural Hallucinogens." *Botanical Museum Leaflets, Harvard University* 28 (2): 123–214.

Schultes, Richard Evans, and Albert Hofmann. 1979. *Plants of the Gods: Origins of Hallucinogenic Use.* New York: McGraw-Hill.

Schultes, Richard Evans, and Siri Von Reis. 1995. *Ethnobotany: Evolution of a Discipline.* Portland, OR: Dioscorides Press.

Sciubba, Jennifer Dabbs. 2021. "Population, Climate, and Conflict: New Data Point to Greater Challenges Ahead," *Center for Strategic and International Studies.* October: https://www.csis.org/analysis/population-climate-and-conflict-new-data-point-greater-challenges-ahead, accessed 31 March 2024.

Sears, John F. 1989. Sacred Places: American Tourist Attractions in the Nineteenth Century. New York: Oxford University Press.

Sea Shepherd Conservation Society. 2019. *In His Own Words: Captain Paul Watson.* https://www.youtube.com/watch?v=YYZnaSr3TII.

Sen, Amartya. 1999. *Development as Freedom.* Oxford: Oxford University Press.

Sheldrake, Merlin. 2023. *Entangled Life: The Illustrated Edition: How Fungi Make Our Worlds.* New York: Random House.

Shepard, Paul. 1967. *Man in the Landscape: A Historic View of the Esthetics of Nature*. New York: Knopf. Distributed by Random House.
———. 1973. *The Tender Carnivore and the Sacred Game*. New York: Scribner.
———. 1982. *Nature and Madness*. San Francisco: Sierra Club Books.
———. 1996. *The Others: How Animals Made Us Human*. Washington, D.C.: Island Press.
Shepard, Paul, and Florence R Shepard. 1998. *Coming Home to the Pleistocene*. Washington, DC: Island Press.
Shiva, Vandana. 1993. "Monocultures of the Mind—Understanding the Threats to Biological and Cultural Diversity." *Indian Journal of Public Administration* 39 (3): 237–48.
———. 1994. *Monocultures of the Mind: Understanding the Threats to Biological and Cultural Diversity*. Hopper Lecture. Guelph, Ont.: Centre for International Programs, University of Guelph.
———. 1997. *Biopiracy: The Plunder of the Nature and Knowledge*. Boston, MA: South End Press.
———. 2002. *Earth Democracy, Living Democracy: Rebuilding Security in an Age of Insecurity*. New Delhi, India: Navdanya : Research Foundation for Science, Technology and Ecology.
Shiva, Vandana. 2005. *Earth Democracy: Justice, Sustainability, and Peace*. Cambridge: South End Press.
———. 2016. *The Violence of the Green Revolution: Third World Agriculture, Ecology, and Politics*. Lexington: University Press of Kentucky.
Silverman, Emily Leah, Dirk Von Der Horst, and Whitney Bauman. 2012. *Voices of Feminist Liberation*. New York: Routledge.
———. 1987. *Religion and the Western Mind*. Pbk ed. Albany, NY: State University of New York Press.
Smart, Ninian. 1996. *Dimensions of the Sacred: An Anatomy of the World's Beliefs*. London, Berkeley: University of California Press.
———. 2008. *The World's Religions*. New York: Cambridge University Press.
Smil, Vaclav. 2008a. *Global Catastrophes and Trends: The next 50 Years*. Cambridge, MA: MIT Press.
———. 2008b. *Oil: A Beginner's Guide*. Oneworld Beginners' Guides. Oxford: Oneworld Publications.
———. 2010. *Energy Transitions: History, Requirements, Prospects*. Santa Barbara, CA: Praeger.
Smith, Jonathan Z. 1982. *Imagining Religion: From Babylon to Jonestown*. Chicago: University of Chicago Press.
———. 1998. "Critical Terms for Religious Studies." In *Critical Terms for Religious Studies*, edited by Mark C. Taylor, 269–284. Chicago: University of Chicago Press.
Snyder, Gary. 1974. *Turtle Island*. New York: New Directions.
Sobrino, Jon. 1994a. *The Principle of Mercy: Taking the Crucified People from the Cross*. Maryknoll, NY: Orbis Books.
———. 1994b. "Jesus the Liberator: *A Historical-Theological Reading of Jesus of Nazareth* . Bloomsbury: Bloomsbury Press.
———. 2008. *The Eye of the Needle: No Salvation Outside the Poor: A Utopian-Prophetic Essay*. London: Darton, Longman and Todd.
Sobrino, Jon, and Ignacio Ellacuría. 1990. *Companions of Jesus: The Jesuit Martyrs of El Salvador*. Maryknoll, NY: Orbis Books.

Song, Choan-Seng. 1996. *Jesus, the Crucified People*. 1st Fortress Press edition. Minneapolis: Fortress Press.
Sosis, Richard, and Candace Alcorta. 2003. "Signaling, Solidarity, and The Sacred: The Evolution of Religious Behavior." *Evolutionary Anthropology* 12: 264–74.
Specter, Michael. 2014. "Vandana Shiva's Crusade Against Genetically Modified Crops." The New Yorker. August 18, 2014.
Spence, Mark David. 1999. *Dispossessing the Wilderness: Indian Removal and the Making of the National Parks*. New York: Oxford University Press.
Sponsel, Leslie E. 2012. *Spiritual Ecology: A Quiet Revolution*. Santa Barbara: Praeger.
Stavenhagen, Rodolfo. 1970. *Agrarian Problems and Peasant Movements in Latin America*. New York: Doubleday Anchor.
Styers, Randall. 2004. *Making Magic Religion, Magic, and Science in the Modern World*. New York: Oxford University Press.
Starhawk. 1999. *The Spiral Dance: A Rebirth of the Ancient Religion of the Goddess*. 20th anniversary edition. San Francisco: HarperCollins.
———. 2013. *The Earth Path: Grounding Your Spirit in the Rhythms of Nature*. San Francisco: HarperCollins.
St. John, Graham. 2008. *Victor Turner and Contemporary Cultural Performance*. New York: Berghahn Books.
Stoll, Mark. 2015. *Inherit the Holy Mountain: Religion and the Rise of American Environmentalism*. Oxford: Oxford University Press.
Suzuki, David. 2022. *The Sacred Balance: Rediscovering Our Place in Nature*. 25th Anniversary. Vancouver: Greystone Books.
Swimme, Brian, and Thomas Berry. 1992. *The Universe Story*. New York: HarperCollins.
Tagliabue, Giovanni Molteni. 2021. "Letter Regarding Dr Vandana Shiva's Anti-Scientific and Unethical Stances." *European Scientist* (blog). October 4, 2021.
Takacs, David. 1996. *The Idea of Biodiversity: Philosophies of Paradise*. Baltimore: Johns Hopkins University Press.
Tambiah, S. J. 1969. "Animals Are Good to Think and Good to Prohibit." *Ethnology* 8 (4): 423–459.
Taussig, Michael T. 1980. *The Devil and Commodity Fetishism in South America*. Chapel Hill, NC: University of North Carolina.
———. 1982. "Peasant Economics and the Development of Capitalist Agriculture in the Cauca Valley, Colombia." *Latin American Perspective*, 62–91.5(3): 62–91.
———. 1987. *Shamanism, a Study in Colonialism, and Terror and the Wild Man Healing*. Chicago: University of Chicago Press.
Taylor, Bron Raymond. 1995. *Ecological Resistance Movements: The Global Emergence of Radical and Popular Environmentalism*. SUNY Series in International Environmental Policy and Theory. Albany: State University of New York Press.
———. 2000. "Deep Ecology and Its Social Philosophy." In *Beneath the Surface: Critical Essays on the Philosohpy of Deep Ecology*, edited by Andrew Light and David Rothenburg. Cambridge: MIT Press.
———. 2002. "Diggers, Wolves, Ents, Elves and Expanding Universes: Bricolage, Religion, and Violence from Earth First! And the Earth Liberation Front to the Antiglobalization Resistance." In *The Cultic Milieu: Oppositional Subcultures in*

an Age of Globalization, edited by Jeffrey Kaplan and Helen Loow. Walnut Creek: Alta Mira Press.

———. 2005a. "Introduction to the Encyclopedia of Religion and Nature." In *Encyclopedia of Religion and Nature*, edited by Bron Taylor London: Continuum

———. 2005b. "Critical Perspectives on the Religions of the World and Ecology." In *Encyclopedia of Religion and Nature*, edited by Bron Taylor: 1375–1376. London: Continuum.

———. 2005c. "Radical Environmentalism." In *Encyclopedia of Religion and Nature*, edited by Bron Taylor:1326–1335. London: Continuum.

———. 2007. "Exploring Religion, Nature, and Culture—Introducing the Journal for the Study of Religion, Nature, and Culture. *Journal for the Study of Religion, Nature, and Culture* 1(1): 5–24.

———. 2010. *Dark Green Religion: Nature Spirituality and the Planetary Future.* Berkeley: University of California Press.

———. 2013. "Kenya's Greenbelt Movement: Contributions, Conflicts, Contradictions, and Complications in a Prominent ENGO." In Tragardh, Lars, Nina Witozek, & Bron Taylor. *Civil Society in an Age of Monitory Democracy.* New York: Berghann Books.

Taylor, Bron. 2016. "The Greening of Religion Hypothesis (Part One): From Lynn White, Jr. and Claims That Religions Can Promote Environmentally Destructive Attitudes and Behaviors to Assertions They Are Becoming Environmentally Friendly." *Journal for the Study of Religion, Nature, and Culture* 10 (3): 268–305.

Taylor, Bron, Gretel Van Wieren, and Bernard Daley Zaleha. 2016. "The Greening of Religion Hypothesis (Part Two) Assessing the Data from Lynn White, Jr. to Pope Francis." *Journal for the Study of Religion, Nature and Culture* 10 (3): 306–78.

Taylor, Bron and Lucas Johnston 2016. "Religion and the Rise of Environmental Politics in the Twentieth Century." In *Wiley-Blackwell Companion to Reliigon and Politics in America.* Oxford: Wiley-Blackwell, 351–368.

Taves, Ann. 2011. *Religious Experience Reconsidered: A Building-Block Approach to Religion and Other Special Things.* Princeton: Princeton University Press.

Taves, Ann, and Michael Kinsella, eds. 2018. "Ethnographies of 'Worldviews' or 'Ways of Life.'" *Religions* 9 (9): 261–279.

Thayer, Robert L. 2003. *LifePlace: Bioregional Thought and Practice.* Berkeley: University of California Press.

Theide, Brian, Matthew Hancock, Ahmed Kodouda. & James Piazza. 2020. "Exposure to Armed Conflict and Fertility in Sub-Saharan Africa." *Demography* 57 (6): 2113–2141.

Tirosh-Samuelson, Hava. 2002. *Judaism and Ecology: Created World and Revealed Word.* Cambridge, Mass: Distributed by Harvard University Press for the Center for the Study of World Religions, Harvard Divinity School.

Tokar, Brian. 1999. *Earth for Sale: Reclaiming Ecology in the Age of Corporate Greenwash.* Boston, MA: South End Press.

Tuan, Yi-Fu. 1974. *Topophilia: A Study of Environmental Perception, Attitudes, and Values.* Englewood Cliffs, NJ: Prentice-Hall.

Tuan, Yi-Fu. 2001. *Space and Place: The Perspective of Experience.* Reprint edition. Minneapolis, MN: University of Minnesota Press.

Tucker, Mary Evelyn. 2006. "Religion and Ecology." In *The Oxford Handbook of Religion and Ecology*, edited by Roger S. Gottlieb. New York: Oxford University Press.

Tucker, Mary Evelyn, and John Grim. 1994. *Worldviews and Ecology: Religion, Philosophy, and the Environment*. Ecology and Justice. Maryknoll, NY: Orbis Books.

Tucker, Mary Evelyn, and Duncan Ryūken Williams. 1997. *Buddhism and Ecology: The Interconnection of Dharma and Deeds*. Cambridge, MA: Harvard University Center for the Study of World Religions, Distributed by Harvard University Press.

Tucker, Mary Evelyn., and John H. Berthrong. 1998. *Confucianism and Ecology: The Interrelation of Heaven, Earth, and Humans*. Cambridge, MA: Distributed by Harvard University Press for the Harvard University Center for the Study of World Religions.

Turner, Victor W. 1969. *The Ritual Process: Structure and Anti-Structure*. Chicago: Aldine Pub. Co.

Turner, Frederick Jackson. 1920. *The Frontier in American History*. Huntington, N.Y. : R. E. Krieger Pub. Co.

Turner, Victor. 1982. *From Ritual to Theatre: The Human Seriousness of Play*. New York: Performing Arts Journal Publications.

Tylor, E.B. 1883. *Primitive Culture*. New York: Henry Holt and Company.

Van Horn, Gavin. 2011. "The Buzzing, Breathing, Clicking, Clacking, Biting, Stinging, Chirping, Howling Landscape of Religious Studies." In *Inherited Land: The Changing Grounds of Religion and Ecology*. Eugene, OR: Pickwick Publicatins.

Van Wieren, Gretel, and Stephen R. Kellert. 2013. "The Origins of Aesthetic and Spiritual Values in Children's Experience of Nature." *Journal for the Study of Religion, Nature & Culture* 7 (3): 243–264.

Veldman, RobinGlobus, Andrew Szasz, and Randolph Haluza-DeLay. 2014. *How the World's Religions Are Responding to Climate Change: Social Scientific Investigations*. New York: Routledge.

Vilaça, Aparecida, and Robin Wright. 2009. *Native Christians Modes and Effects of Christianity among Indigenous Peoples of the Americas*. Farnham, Surrey, UK: Ashgate.

Von Bertalanffy, Ludwig. 1972. "The History and Status of General Systems Theory." *The Academy of Management Journal* 15 (4): 407–26.

von Stuckrad, Kocku. 2007. "Finding Data: Some Reflections on Ontologies and Normativities." *Journal for the Study of Religion, Nature & Culture* 1 (1): 39–46.

Wadleigh, Michael, Bob Maurice, Richie Havens, Joan Baez, Joe Cocker, Joe McDonald, Arlo Guthrie, et al., dirs. 2009. *Woodstock 3 days of peace & music*. Burbank, CA: Warner Bros. Pictures : Distributed by Warner Home Video.

Waldau, Paul. 2003. *The Specter of Speciesism: Buddhist and Christian Views of Animals*. American Academy of Religion Academy Series. New York: Oxford University Press.

———. 2010. *Animal Rights: What Everyone Needs to Know*. New York: Oxford University Press.

———. 2013. *Animal Studies: An Introduction*. New York: Oxford University Press.

Waldau, Paul, and Kimberley C. Patton. 2006. *A Communion of Subjects: Animals in Religion, Science, and Ethics*. New York: Columbia University Press.

Weatherdon, Meaghan Sarah. 2017. "Indigenous Knowledge and Contested Spirituality in Canadian Nuclear Waste Management." *Journal for the Study of Religion, Nature and Culture* 11 (1): 86–108.

Weaver, Jace. 1996. *Defending Mother Earth: Native American Perspectives on Environmental Justice*. Maryknoll, NY.: Orbis Books.

Weber, Max. 2005. *The Protestant Ethic and the Spirit of Capitalism*. London and New York: Routledge.

White, Jr., Lynn. 1967. "The Histocial Roots of Our Ecologic Crisis." *Science* 155 (3767): 1203–1207.

Whitehead, Neil L., and Robin Wright, eds. 2004. *In Darkness and Secrecy: The Anthropology of Assault Sorcery and Witchcraft in Amazonia*. Durham, NC: Duke University Press.

Wiethaus, Ulrike. 2013. *The Seven Rites of the Lakota*. CreateSpace: CreateSpace.

Wilkinson, Loren, ed. 1980. *Earthkeeping: Christian Stewardship of Natural Resources*. Grand Rapids: William B. Eerdmans.

Wilson, David Sloan. 2002. *Darwin's Cathedral: Evolution, Religion, and the Nature of Society*. Chicago: University of Chicago Press.

Wilson, Edward Osborne. 1975. *Sociobiology*. Abridged ed. Cambridge: Belknap Press of Harvard University Press.

Wilson, Edward O. 1984. *Biophilia*. Cambridge, Mass: Harvard University Press.

Winkelman, Michael. 2009. "Shamanism and the Origins of Spirituality and Ritual Healing." *Journal for the Study of Rleligion, Nature, and Culture* 3 (4): 458–89.

Witt, Joseph, and Bron Taylor. 2017. "Special Issue Introduction: Religion and Eco-Resistance Movements in the Twenty-First Century." *Journal for the Study of Religion, Nature and Culture* 11 (1): 5–22.

Wittgenstein, Ludwig. 1968. *Philosophical Investigations*. 3rd edition. Oxford: Blackwell.

World Commission on Environment and Development, (WCED). 1987. *Our Common Future*. Oxford: Oxford University Press.

Worster, Donald. 1979. *Nature's Economy: A History of Ecological Ideas*. Cambridge; New York, NY, USA: Cambridge University Press.

———. 1993. *The Wealth of Nature: Environmental History and the Ecological Imagination*. New York: Oxford University Press.

Worthy, Kenneth, Elizabeth Allison, and Whitney Bauman. 2019a. *After the Death of Nature: Carolyn Merchant and the Future of Human-Nature Relations*. New York: Routledge.

Wright, Robin. 1988. "Anthropological Presuppositions of Indigenous Advocacy." *Annual Review of Anthropology* 17: 365–90.

Wright, Robin. 2009. "The Art of Being Crente: The Baniwa Protestant Ethic and the Spirit of Sustainable Development." *Identities: Global Studies in Culture and Power* 16 (2): 202–226.

Worster, Donald. 1979. *Nature's Economy: A History of Ecological Ideas*. Cambridge; New York, NY, USA: Cambridge University Press.

———. 1993. *The Wealth of Nature: Environmental History and the Ecological Imagination*. New York: Oxford University Press.

———. 1993. *The Wealth of Nature: Environmental History and the Ecological Imagination*. New York: Oxford University Press.

Wright, Robin. 1988. Anthropological Presuppositions of Indigenous Advocacy. *Annual Review of Anthropology* 17: 365–390.

Wright Robin. 2009. "The Art of Being Crente: The Baniwa Protestant Ethic and the Spirit of Sustainable Development." *Identities: Global Studies in Culture and Power* 16 (2): 202–226.

York, Michael. 2006. "Magic." In *The Encyclopedia of Religion and Nature*. Continuum.

INDEX

Note: *Italic* page numbers refer to *figures* and page numbers followed by 'n' denote notes.